B. Eng. Axel Jörn

Programmierung eines Mikrocontrollers

Jörn, B. Eng. Axel: Programmierung eines Mikrocontrollers. Hamburg, Bachelor + Master Publishing 2015
Originaltitel der Arbeit: Programmierung eines Mikrocontrollers

Buch-ISBN: 978-3-95820-326-6
PDF-eBook-ISBN: 978-3-95820-826-1
Druck/Herstellung: Bachelor + Master Publishing, Hamburg, 2015
Covermotiv: © Kobes · Fotolia.com
Zugl. bbw Hochschule, Berlin, Deutschland, Studienarbeit, September 2012

Bibliografische Information der Deutschen Nationalbibliothek:
Die Deutsche Nationalbibliothek verzeichnet diese Publikation in der Deutschen Nationalbibliografie; detaillierte bibliografische Daten sind im Internet über http://dnb.d-nb.de abrufbar.

© Bachelor + Master Publishing, Imprint der Diplomica Verlag GmbH
Hermannstal 119k, 22119 Hamburg
http://www.diplomica-verlag.de, Hamburg 2015
Printed in Germany

Inhalt

Abbildungsverzeichnis und Tabellenverzeichnis

Abkürzungsverzeichnis

µC	Mikrocontroller/Mikrocomputer/MCUs
SD	Secure Digital
ALU	Arithmetic Logic Unit
RAM	Random Access Memory
CPU	Central Processing Unit
RISC	Reduce Instruction Set Computer
CISC	Complex Instruction Set Computer
USB	Universal Serial Bus
COM	Component Object Model
IDE	Silicon Laboratories Integrated Development Environment
IR	Interruptfrequenz
RLV	Reloadvalue
ISR	Interrupt Service Routine

Gleichungsverzeichnis

1 Einleitung

1.1 Allgemeines

Die vorliegende Arbeit entstand im Rahmen der „Student Consulting" an der Northern Business School im Wintersemester 2011/2012 im Fach Informations- und Entwurfsmethoden für den Studiengang Bachelor of Engineering Maschinenbau mit Mechatronik. Ein besonderer Dank gilt meinem einfallsreichen, hilfsbereiten und engagierten Dozenten Prof. Dr. Hahlweg.

1.2 Zielsetzung

Die Zielsetzung des Projektes ist es, einen Mikrocontroller "C8051 F0200" von „Silicon Laboratories" zu programmieren und dabei das Hochzählen im Sekundentakt zu realisieren. Die Programmierung des Mikrocontrollers kann sowohl mit „Visual C++ 2010 Express" als auch mit der beigelegten Development Software IDE auf „C-Basis" durchgeführt werden. Der dabei entstehende Programmcode soll charakterisiert und anschließend auf einer 7-Segmentanzeige ausgegeben werden. Diese Arbeit stellt im weiteren Sinne eine Dokumentation der einzelnen Entwurfsschritte da und möge die Funktionalität der dabei entstehenden Digitaluhr beweisen. Dabei ist vorrangig auf die Möglichkeit der Umsetzung im Labormaßstab bei minimalen Kosten zu achten.

2 Grundlagen

2.1 Theoretische Grundlagen

Im ersten Teil dieser Arbeit sollen die für die angestrebte Mikrocontrollerprogrammierung herangezogenen theoretischen Grundlagen vorgestellt werden. Dabei wird lediglich auf die allgemeine Funktionssystematik der wichtigsten Teile eines Mikrocontrollers eingegangen und nicht auf Zahlensysteme, Codierungen oder Bits & Bytes. Auch die einzelnen Module der C-Programmierung (wie z.B.: Verzweigungen, Schleifen, Funktionen, Variablen und Konstanten) können nicht tiefer behandelt werden. Vielmehr soll der Umgang mit einigen von ihnen im Praxisteil vorgestellt werden. Der Theorieteil wird sehr kurz gehalten, da eine tiefere Betrachtung den Rahmen dieser Arbeit sprengen würde und zu schnell veraltet. Die Abbildung auf der folgenden Seite zeigt den wesentlichen Aufbau eines Mikrocontrollers. Anhand der aufgeführten Bauteile werden einige wichtige Funktionen kurz beschrieben.

2.1.1 Überblick über die Mikrocontroller

Mikrocontroller findet man heute in fast jedem elektronischen Gerät. Sie kommen in Thermometern zur Anzeige der Temperatur und in Steuereinheiten für Kaffeeautomaten zum Einsatz. Sie öffnen und schließen Garagentore und unterstützen den Autofahrer durch ABS und ESP. Ein Mikrocontroller[1] oder auch Mikrocomputer ist ein elektronischer Baustein, der Rechnungen und Steuerungen ähnlich wie ein PC ausführen kann. Allerdings hat dieser lediglich ein Gerät zu steuern, während ein PC flexibler ist und viele unterschiedliche Geräte steuern und auswerten muss. Da Mikrocontroller nur für eine spezielle Aufgabe verantwortlich sind, hat dieser im Vergleich zum PC eine sehr geringe Rechenleistung und wenig Speicher. Die Grenzen zwischen µC und ein PC sind allerdings fließend. Ein gutes Beispiel dafür ist ein Mobiltelefon. In einem Mobiltelefon wird ein Mikrocontroller zum Wählen der

---

[1] Auch oft als µC bezeichnet.

Rufnummer und den Aufbau des Telefonats benötigt. Mit Handys kann man heute aber auch Videos machen und Musik hören. Für diese Funktionen bedarf es allerdings schon einens PC. Da es nicht für jede Aufgabe einen speziellen Mikrocomputer gibt, kann man aus einer Vielzahl von Baumustern einen geeigneten selektieren. Es gibt unterschiedliche Hersteller, die viele Varianten von Controllern produzieren, welche sich aber in der grundsätzlichen Funktionsweise kaum unterscheiden. Mikrocontroller variieren in der Anzahl von Ein- und Ausgängen sowie über spezielle Hardwarebausteine im Inneren, mit denen verschiedene Aufgaben vereinfacht werden. In der Regel verfügen µCs über einen Timer, mit dem man Zeiten bestimmen und Signale definierter Länge generieren kann. Außerdem haben viele Bausteine einen integrierten Analog-Digital-Wandler, der es ermöglicht, analoge Signale wie z.B. Temperatur oder Batteriespannung zu messen. Die Große des Mikrocontrollers wird durch die Anzahl seiner Pins definiert. Die Anzahl der Pins ist vom Aufgabengebiet des Kontrollers abhängig. Es darf festgehalten werden, dass auch die Rechenleistung mit der Anzahl der Pins steigt, da bei vielen Pins auch mehr Aufgaben bewältigt werden müssen. Auch die Anzahl der Bits, die gleichzeitig verarbeitet werden, steigt mit der Größe des Mikrocontrollers. Es gibt Baueinheiten mit einer Busbreite von 4 Bit – 64 Bit. Am weitesten verbreitet sind µCs mit 8 bis 16 Bit. 32 Bit Mikrocomputer sind meist in Geräten eingesetzt, die einen Farbdisplay ansteuern oder zum Datenaustausch mit dem PC verbunden (SD Karten) werden. Für die meisten Anwendungen sind günstige 8-Bit-Mikrocontroller ausreichend. [URL 1] [MFE, 2009]

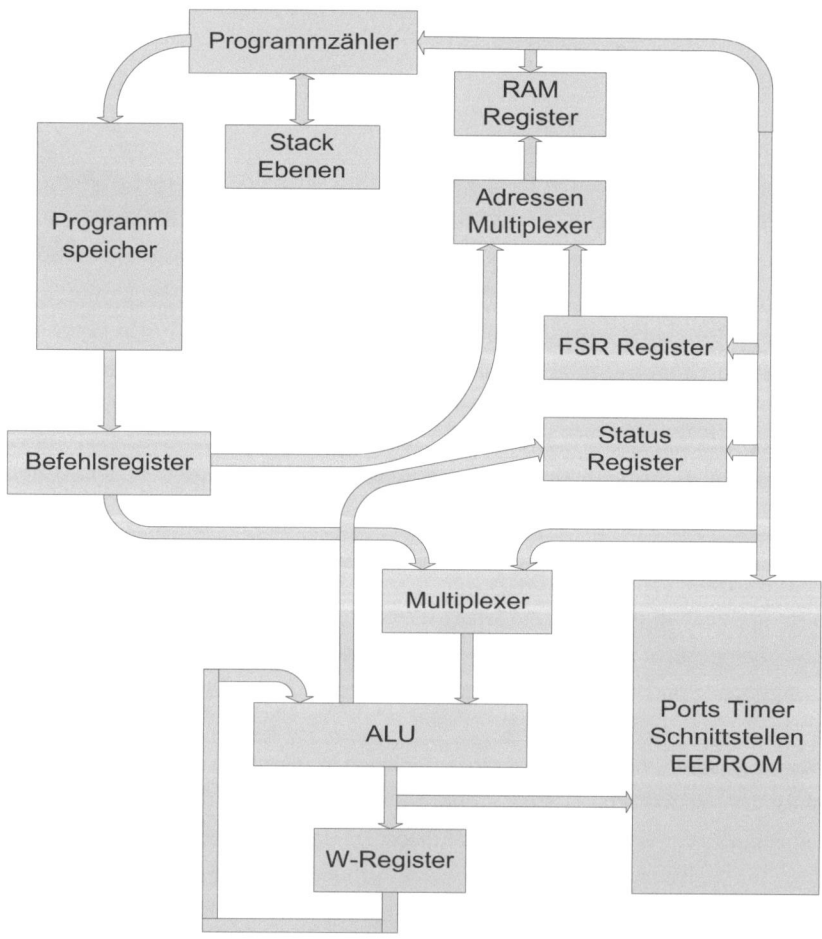

Abbildung 1: Allgemeines Blockdiagramm eines Mikrocontrollers [MFE, 2009]

2.1.2 Die Arithmetic Logic Unit ALU (Rechenwerk, Operationswerk)

Damit ein µC überhaupt funktioniert, sind viele Funktionsblöcke nötig, die optional zusammenarbeiten müssen. Die Arithmetic Logic Unit ist dabei der wichtigste Bestandteil. Sie ist das Regelwerk des Controllers. In dieser Einheit werden Addition und weitere logische Operationen durchgeführt. Damit sie mit den richtigen Werten rechnen kann, müssen diese zum richtigen Zeitpunkt vom richtigen Speicherort zur Verfügung gestellt werden. [MFE, 2009] [MT, 2009]

2.1.3 Der Programmspeicher

Zu Beginn muss der Mikrocontroller mit Daten versorgt werden. Der µC versteht keine Programmiersprache wie Basic, C oder Assembler und muss daher mit rein binären Werten (also Nullen und Einsen) beschrieben werden. Diese werden dann im Programmspeicher beginnend bei der Adresse 0x0000 abgelegt. Da der Speicher ein nicht flüchtiger Speicher ist, der die Daten auch nach dem Ausschalten behält, wird hier der Programmcode gespeichert. Läge man die Daten in den RAM-Speicher, würde das Gerät nach dem Ausschalten nicht mehr funktionieren. Um nun Daten zu generieren, die der PIC versteht, sind einige Schritte nötig. Theoretisch könnte man direkt den Programmcode in Form von binären Werten schreiben und mit einem Programmiergerät in den Mikrocontroller laden. Dieses Vorgehen wäre jedoch impraktikabel und viel zu aufwändig. Daher werden verschiedene Programmiersprachen zur Mikrocontroller Programmierung eingesetzt. Gängige sind C oder Assembler. Beide Sprachen sind hardwarenah und können daher direkt mit den physikalisch vorhandenen Registern kommunizieren. Für größere Mikrocontroller wird fast ausschließlich C verwendet; bei kleineren häufig Assembler, um die Hardware optimal auszunutzen. Für unser Projekt werden wir die C-Programmierung einsetzen. [MFE, 2009] [ITG1, 2010] [MT, 2009]

2.1.4 Der Compiler und der Linker

Wie bereits erwähnt, wäre es viel zu umständlich, in Maschinensprache[2] zu programmieren. Daher wird ein Programm, welches den C-Code oder den Assembler-Code enthält, in Maschinensprache übersetzt. Man nennt dieses Tool Compiler. Er geht Schritt für Schritt durch den Code und interpretiert die vom Entwickler programmierten Befehle. Dies geschieht häufig in mehreren Durchgängen. Nach dem Kompilieren müssen die einzelnen Module mit einem weiteren Tool, dem Linker, verbunden werden. Dieser liefert den Maschinencode, welcher auf dem vorgesehenen Controller funktionsfähig ist. Während des Übersetzungsvorgangs wird eine Datei mit der Endung „.hex" generiert, die beschreibt, an welcher Stelle im Speicher die einzelnen Befehle stehen sollen. Dieses File kann anschließend mit einem Programmiergerät in den Mikrocontroller geladen werden. Dadurch gelangt das Programm in den Programmspeicher. Durch einen internen oder externen Takt wird der Programmzähler schrittweise hochgezählt und führt so einen Befehl nach dem anderen aus. Die Befehle werden dann im Befehlsregister verarbeitet und die Daten und RAM-Adressen werden über die internen Leitungen an das Rechenwerk (ALU) oder den RAM-Speicher weitergegeben. [MFE, 2009] [ITG1, 2010] [MT, 2009] [ITH, 2005]

2.1.5 Die Datenverarbeitung in der ALU

Nachdem die Daten in der ALU (dem Rechenwerk) angekommen sind, können sie nun verarbeitet werden. Dabei werden die Daten zunächst in das W-Register geladen und anschließend in das RAM-File Register weitergeleitet. Die ALU kann nur einfache logische Operationen ausführen, daher sind alle Befehle logisch verknüpft. Es ist möglich die Register durch UND(AND), ODER (OR) und XODER

---

[2] Ist die einzige Sprache, die ein Prozessor wirklich versteht. Ausführbare Programme bestehen aus Maschinencode. Die Kombinationen der einzelnen Bits steuert die CPU.

3

(XOR) zu verknüpfen. Die Daten können also nur addiert und subtrahiert werden, eine Multiplikation oder Division ist nicht möglich und muss durch die Kombination der Grundbefehle aufgebaut werden. [MFE, 2009] [ITH, 2005]

2.1.6 Das Statusregister

Um den Status der einzelnen Befehle zu erhalten, markiert das Statusregister diese mit sogenannten Statusflags (Status-Flaggen). Diese zeigen an, was genau bei einer Operation geschehen ist. Unter weiteren handelt es sich dabei z.B. um das Carry-Flag (C), das Digit-Carry-Flag (DC) und das Zero-Flag (Z). Das Carry-Flag zeigt einen Übertrag an, wenn beispielsweise nach einer Addition, das Ergebnis nicht mehr mit 8 Bit dargestellt werden kann. Das Digit-Carry Flag zeigt einen Übertrag nach dem vierten Bit an. Die Zero-Falg zeigt an ob, nach einer Operation das Ergebnis Null ist. [MFE, 2009] [MT, 2009]

2.1.7 Die Assemblerbefehle und Befehlsübersicht

Die meisten µC sind RISC-Prozessoren. RISC steht für Reduce Instruction Set Computer und bedeutet übersetzt „Computer mit reduzierten Befehlen". Das heißt, dass der Controller mit einem begrenzten Befehlssatz auskommt. Das Gegenteil ist der CISC-Prozessor. CISC heißt „Complex Instruction Set Computer". Er verfügt über einen komplexeren Befehlsumfang. Dabei werden mehrere Befehle zu einem Befehl zusammengefasst. Der Vorteil von CISC-Prozessoren ist eine vereinfachte Programmierung. Allerdings müssen dazu mehr Befehle erlernt werden und jeder Befehl hat je nach Umfang eine unterschiedlich lange Ausführungsdauer von meist mehreren Befehlstakten. [MFE, 2009] [8051]

Der Mikrocontroller „C8051 F0200" von Silicon LABS ist ein 8-Bit-RISC-Prozessor der Harvard Architektur; sein entsprechender Befehlssatz kann dem „8051 Instruction Set"[3] entnommen werden. Weitere theoretische Grundlagen finden sich z.B. in den Werken „Mikrocontroller für Einsteiger", „Mikroprozessortechnik" und „Die Mikrocontroller 8051, 8052 und 80C517" aus dem Literaturverzeichnis. Die theoretischen Grundlagen zum zu unserem „C8051 F0200 Development Kit" sind unter http://www.silabs.com/Support%20Documents/Software/8051_Instruction_Set.pdf zu finden. In diesem Dokument soll aufgrund der zunehmenden Komplexität und limitierten Umfang der SCA nicht weiter darauf eingegangen werden. Vielmehr soll die Handhabung und das Verhalten des „C8051 F0200" von Silicon LABS bei Programmierung mit C beschrieben werden. [URL 2] [8051]

2.2 Der Stand der Mikrocontrollertechnik

Um den aktuellen Stand der Technik darzustellen, bietet sich die Recherche im Internet an. Einen sehr guten Überblick und eine große Auswahl an Microcomputern bietet die Seite: http://www.mikrocontroller.net und https://www.mikrocontroller.com Hier findet sich eine Vielzahl selbiger von 4Bit -64 Bit mit den dazugehörigen Datenblättern, Kaufpreisen und Lieferanten. Auch die folgenden Internetseiten http://elmicro.com/ und http://www.silabs.com bieten eine Vielzahl von Produkten und Informationen zu diesem Thema.

Übungs- und Programmierbeispiele für den „8051" sind auf http://vdf.ethz.ch erhältlich.

Auf den Internetseiten von Pearl oder Conrad finden sich Lern- und Anwendungspakete zum Thema Mikrocontrollertechnologie.

http://www.conrad.de/ce/ and http://www.pearl.de/c.shtml

---

[3] Vgl. URL 3: 8051 Instruction Set

3 Die Mikrocontrollerprogrammierung in der Praxis

3.1 Die benötigten Ressourcen

Für die Umsetzung der Aufgabe ist folgende Hardware und Software notwendig: Mikrocontroller (C8051 F0200 Development Kit) mit dem dazugehörigen Treiber, USB-Debugger, Kabel und einen Laptop oder PC (z.B. Asus Notebook N5 Series), auf welchem eine C-Programmiersoftware lauffähig ist, sowie eine C-Programmiersoftware (z.B. Visual C++ 2010 Express). Die folgende Abbildung zeigt das Blockdiagramm für den inneren Aufbau des Cip-51-Mikrocontrollers und C8051-F0200 Development Kits). Um die Digitaluhr anzeigen zu können, muss eine kleine „Schaltung" aufgebaut werden. Hierzu wird eine 7-Segmentanzeige (Hewlett Packard 5082-7740), 75Ω-Widerstände, Breitbandkabel mit Stecker und eine Steckbrett benötigt. Um den Aufbau umzusetzen, werden die typischen Elektronick-Werkzeuge wie Lötkolben, Seitenschneider usw. benötigt.

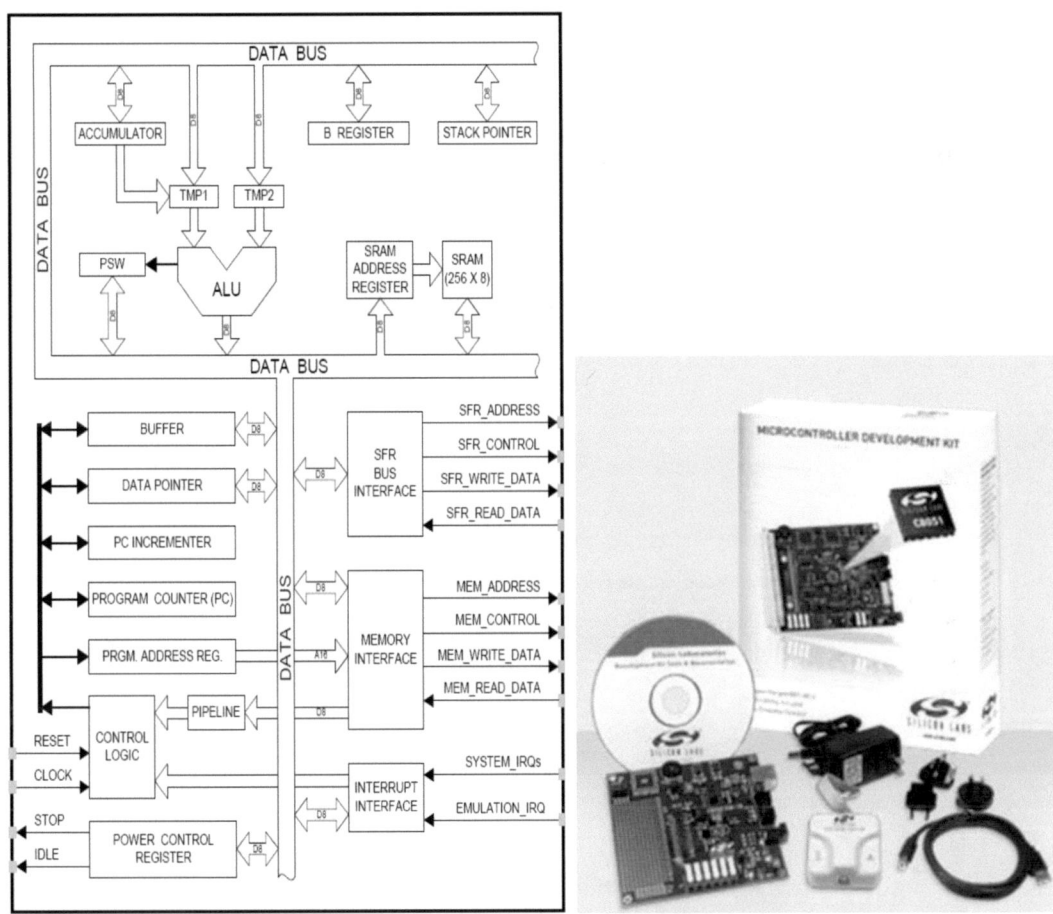

Abbildung 2: Blockdiagramm des CIP-51-Mikrocontrollers und C8051F0200 Development Kit.

3.2 Der allgemeine Versuchsaufbau

Der allgemeine Versuchsaufbau besteht lediglich darin, dass der Mikrocontroller über den USB Debugger mit dem Rechner verbunden wird. Nun muss für eine funktionierende Kommunikation zwischen den beiden Geräten gesorgt werden. Dafür müssen diverse Treiber[4] installiert werden. Eine leicht verständliche Beschreibung hierfür ist im Development Kit vorhanden. Dann wird die kostenlose Programmiersoftware „Visual C++ 2010 Express" installiert und gestartet. Diese kann man als 30 Tage Testversion von der Microsoft-Homepage herunterladen und nach Ablauf der 30 Tage für ein Jahr online freischalten lassen. Um letzteres zu testen, wird ein neues Projekt (Name Test) angelegt, welches auf einer „Win32-Konsolenanwendung[5]" basiert. Nach dem Speichern haben wir bereits ein lauffähiges Programm, das wir in folgendem nach unseren Bedürfnissen und Aufgaben durch Programmierung charakterisieren können. Die erste Abbildung stellt den Versuchsaufbau dar. Die zweite Abbildung zeigt eine Win32-Konsolenanwendung in „Visual C++ 2010 Express".

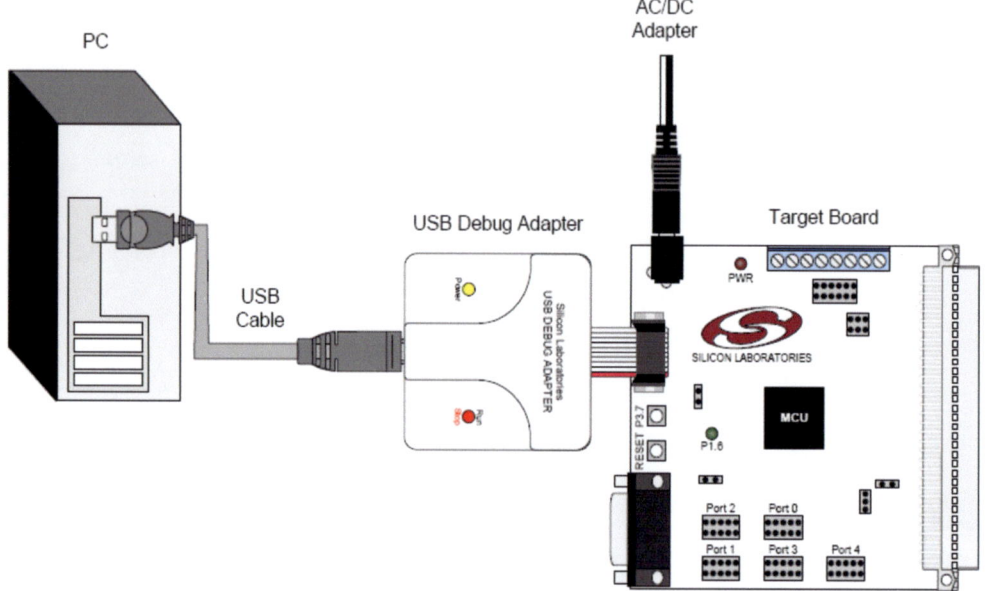

Abbildung 3: Der allgemeine Versuchsaufbau [Anhang 2]

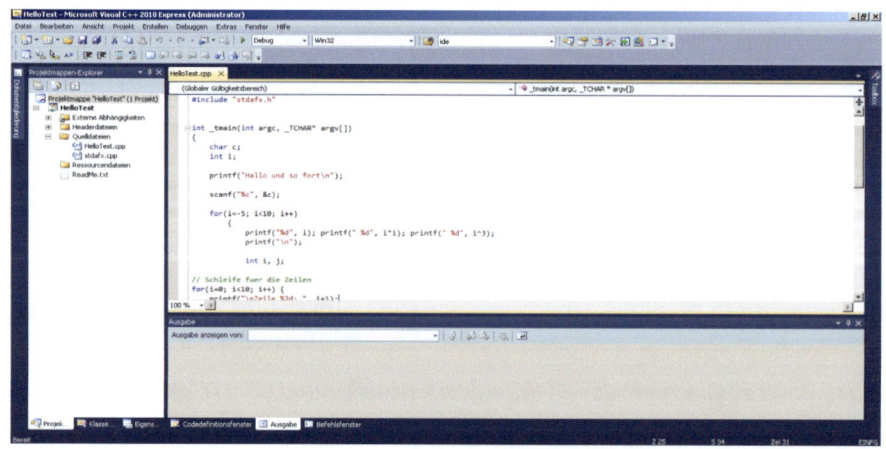

Abbildung 4: Win32-Konsolenanwendung in „Visual C++ 2010 Express"

---

[4] Die Silicon Laboratories Integrated Development Environment (IDE) ist die Entwicklungsumgebung. Die Keil 8051 Development Tools beinhalten den macroassembler, linker, evaluation 'C' compiler.
[5] Ist ein Computerprogramm ohne grafische Benutzeroberfläche, welches lediglich über Textkommandos gesteuert wird. (*Graphical User Interface* oder GUI)

3.2.1 Der Test des Versuchsaufbaus mit „Blinky stripped"

Um die reibungslose Kommunikation zwischen den fünf Komponenten (Rechner, C-Programm, Treiber USB-Debugger und µC) zu prüfen, wurde in die bereits installierte „Silicon Laboratories Integrated Development Environment [IDE]" das mitgelieferte Beispielprojekt „Blinky" geöffnet. Ein auf das Wesentliche reduzierter Programmcode „Blinky_stripped" findet sich im Anhang 3. Dieser lässt die auf dem Target Board installierte grüne LED in einer bestimmten Frequenz blinken. Als erstes muss das Programm auf den Chip übertragen werden und kann dann mit der „Play" Taste ausgeführt werden. Die LED blinkt mit einer Frequenz von fünfmal pro Sekunde. Mit diesem Projekt werden verschiedene Basisfunktionen des „C8051" getestet. Dies sind:

- disabling the watchdog timer (WDT),
- configuring the Port I/O crossbar,
- configuring a timer for an interrupt routine,
- initializing the system clock,
- configuring a GPIO port.

Nun kann man das USB-Kabel von dem Target Board abziehen und die LED blinkt weiter. Das bedeutet, dass das Programm selbstständig auf dem µC weiterläuft. Mit diesem Test und der lauffähigen „Win32-Konsolenanwendung" sind die erforderlichen Grundlagen zur Umsetzung des Projektes gegeben.

3.2.2 Weitere Tests und Änderung des Programmcodes mit „Blinky stripped"

Um sich mit dem Mikrokontroller und deren Verhalten vertraut zu machen, wurde das C-Programm verändert. Unter dem Kommentar „// disabling the watchdog timer" wurde die Blinkfrequenz verändert „Timer3_Init (SYSCLK / 12 / 10);". Für fünfmal blinken pro Sekunde steht die „10" (10/2=5) Um die LED dreimal pro Sekunde blinken zu lassen kann die Zahl auf „6" (6/2=3) geändert werden. Dabei sei erwähnt, dass man eine minimale Blinkfrequenz von ca. 1,5mal pro Sekunde einstellen kann, da man sonst aus dem 16-Bit-Rahmen (2^{16} = 65536 Werte) läuft. Die Spannung, welche die LED zum Blinken bringt, kann über die Pins 1 und 7 von Port 1 auf dem Oszilloskop dargestellt werden. Die folgenden Formeln beschreiben die Beziehung zwischen SysClock=2MHz[6], Reloadvalue(RLV), Takt=12, Interruptfrequenz(IF)=10Hz.

$$RLV = \left(\frac{SysClock}{Takt}\right) : IF \qquad\qquad 3.1$$

Es wird ersichtlich dass, je größer die Interruptfrequenz wird, desto kleiner wird der RLV. In unserem Fall lautet die Rechnung:

$$RLV = \left(\frac{2000000\ Hz}{12\ Takte}\right) : 10Hz = 16666,666\ Werte < 65536\ Werte$$

Als nächstes folgt ein Beispiel mit 2-Hz-Interruptfrequenz, welche aus dem 16Bit Rahmen läuft und daher mit dem 8051 nicht verarbeitet werden kann.

$$RLV = \left(\frac{2000000\ Hz}{12\ Takte}\right) : 2Hz = 83333,333\ Werte > 65536\ Werte$$

Die Spannung, welche die LED zum blinken bringt, kann über die Pins 1 und 7 von Port 1 auf dem Oszilloskop dargestellt werden. Das nächste Foto zeigt dieses „Rechtecksignal" auf einem Digital Oszilloskop.

---

[6] Eigentlich schwankend zwischen 1,6MHz und 2,4MHz.

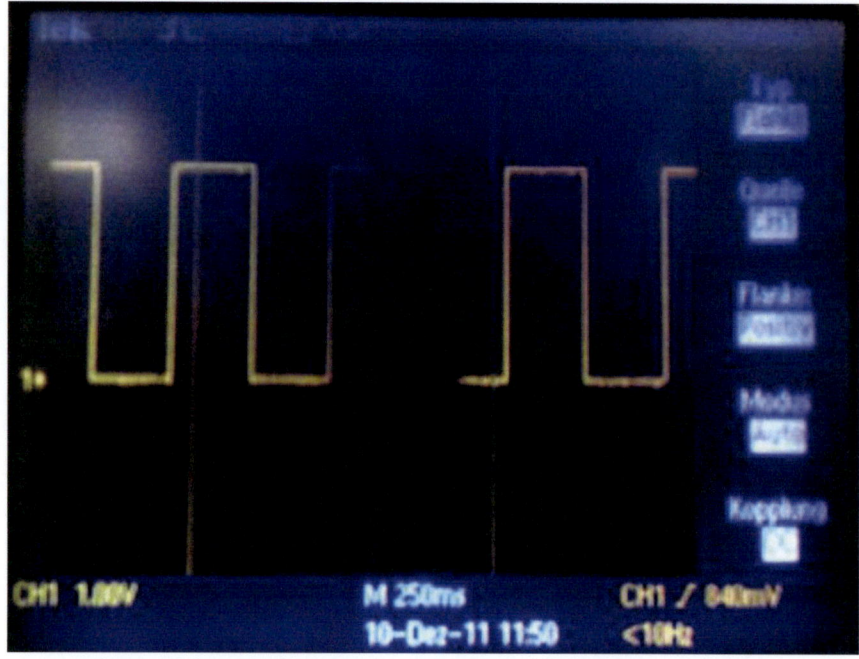

Abbildung 5: Rechtecksignal an grüner LED vom Code "Blinky_stripped"

Auf Basis dieses Verhaltens soll die Uhr programmiert werden. Zunächst wird die Inkrementfrequenz (f_{int}), die sich daraus ergebende Inkrementzeit (T_{int}) und die benötigte Reloadzeit (T_{rel}) berechnet.

$$f_{int} = \frac{Sys\ Clock}{Takt} \qquad\qquad 3.2$$

$$f_{int} = \frac{2000000}{12} = 166666{,}66\ s^{-1}$$

$$T_{int} = \frac{1}{f_{int}} \qquad\qquad 3.3$$

$$T_{int} = \frac{1}{166666{,}66\ s^{-1}} = 0{,}6 * 10^{-5} s$$

$$T_{rel} = T_{int} * 2^{16} \qquad\qquad 3.4$$

$$T_{rel} = 0{,}6 * 10^{-5} * 65536\ Werte = 0{,}3932\ s \approx 0{,}4s$$

3.3 Die Programmierung der Uhr

Die eigentliche Uhrenfunktion ist ein C-Programm, welches auf den Mikrocontroller übertragen wird und von dort aus die einzelnen Segmente der 7-Segmentanzeige ansteuert. In diesem Teil der SCA wird die Inbetriebnahme der 7-Segmentanzeige und der Programmcode beschrieben. Abschließend wird der Funktionstest dargestellt.

3.3.1 Das Anschließen der 7-Segmentanzeige

Zu Beginn muss die Pin-Belegung der Anzeige ermittelt werden. Dabei wurden die einzelnen Pins per Diodentest mit einem Multimeter „abgefragt". Die Eigenspannung des Messgerätes ist ausreichend, um die Dioden zum Leuchten zu bringen und diese ihrem Pin zuzuordnen. Dieser Test lieferte folgendes Ergebnis: die nächste Grafik zeigt an, welche Diode (Balken) mit welchem Pin angesteuert werden kann. Die Tabelle zeigt, welche Dioden Leuchten müssen, um eine bestimmte Zahl anzuzeigen

(Binärcode) und das dazugehörige Bitmuster als Dezimalzahl. Zum Beispiel müssen, um die Zahl sieben anzuzeigen, die Dioden an Pin 3,2,1 angesteuert werden. Dies entspricht dem Bitmuster 1+4+8.

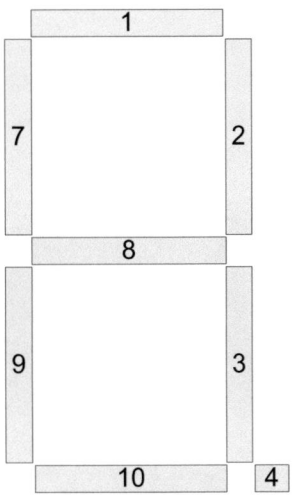

Pin 5&6 = Ground

Zahl	Balkennummer auf 7Segma. (Binärcode)								Duale Aufsummierung (Bitmuster)	Ergebnis: Dezimal
	10	3	2	1	7	9	8	4		
0	1	1	1	1	1	1	0	0	1+2+4+8+16+32	63
1	0	1	1	0	0	0	0	0	2+4	6
2	1	0	1	1	0	1	1	0	1+4+8+32+64	109
3	1	1	1	1	0	0	1	0	1+2+4+8+64	79
4	0	1	1	0	1	0	1	0	2+4+16+64	86
5	1	1	0	1	1	0	1	0	1+2+8+16+64	91
6	1	1	0	1	1	1	1	1	1+2+8+16+32+64+128	251
7	0	1	1	1	0	0	0	0	2+4+8	14
8	1	1	1	1	1	1	1	0	1+2+4+8+16+32+64	127
9	1	1	1	1	1	0	1	1	1+2+4+8+16+64+128	223
Bit:	1	2	4	8	16	32	64	128		

Abbildung 6: Bitmuster für gewünschte Zahlen & Pinzuordnung (Binärcode) der 7-Segmentanzeige.

Im nächsten Schritt wurde die 7-Segmentanzeige auf ein Steckbrett gelötet[7] und die einzelnen Pins über einen 75Ω-Vorwiderstand[8] mit einem Breitbandkabel verbunden. Dabei ist auf die entsprechende Farbe (Pin Belegung auf dem Board) zu achten.[9] Die genaue Pin-Definition findet sich in Ta-

---

[7] Die Leiterbahn zwischen den Pins der Anzeige muss nach dem auflöten mittels einer Feile unterbrochen werden, um die einzelnen Pins nicht kurzzuschließen.
[8] An den roten LEDs der Anzeige fallen 1,5V ab. Um 1,5 abzubauen, werden bei 20mA 75Ω Vorwiderstände benötigt. (1,5V/0,02A=75Ω).
[9] Auf der weißen Phase an Pin 9 liegen konstant 3,3V DC. Da diese nicht benötigt werden, wurde die Phase „weggebunden", um einen möglichen Kurzschluss zu vermeiden.

belle 5 auf Seite 8 im Anhang 2. Das gleiche Vorgehen muss nun für die Anzeige der 10er-Sekunden-Stelle wiederholt werden. Mit den beiden Anzeigen ist es möglich, das Hochzählen von 0 bis 19 Sekunden anzuzeigen. Das Foto soll die Pin-Belegung der Breitbandkabel verdeutlichen. Außerdem wird ersichtlich, mit welchen Ports die Kabel auf dem Target Board verbunden werden. Später bei der Beschreibung des Programmcodes wird deutlich, warum die Sekunden an Port 0 (P0) und die 10er-Sekunden-Stelle an Port 1 (P1) ausgegeben werden. Steckt man eine weitere Anzeige an Port 2 (P2) werden mit dem Programmcode in Anhang 4 die Minuten von 1-9 angezeigt. Insgesamt besitzt der Controller 8051 10pin parallel Ports (0-7) mit je 3,3V Gleichstrom und digitalem Ground.

Abbildung 7: Fertiggestellte 7-Segmentanzeigeschaltung mit Verbindung zum Mikrocontroller

10

3.3.2 Der Programmcode

Als Grundlage für den benötigten Programmcode dient der oben erwähnte C-Code von „Blinky_stripped", welcher im Anhang 3 wiederzufinden ist. Der daraus entwickelte Programmcode für die Uhr liegt im Anhang 4 und wird im Folgendem genauer beschrieben. Als erstes werden unter `// Includes` im Programmcode die Bibliothek und unser 8051-Chip als Zielobjekt eingebunden. Dieser Befehl heißt „#include <c8051f020.h>". Mit den nächsten Zeilen wird definiert, dass zwei Timer benötigt werden.

```
sfr16 TMR3RL  = 0x92;

sfr16 TMR3    = 0x94;.
```

Unter `// Global CONSTANTS` wird zunächst mit dem Befehl SYSCLOK die Abtastfrequenz auf 2 MHz des internen Oszillators festgelegt. Anschließend werden die „Charakter" für die Zeiteinheiten und die Bitmuster in Dezimalzahlen für die anzuzeigenden Zahlen nach der Tabelle in Abbildung 6 definiert.

```
#define SYSCLK 2000000      // 2MHz for OSCICN.[1:0] = 00
                            // approximate SYSCLK frequency in Hz
char   oscmask = 0x00;      // OSCICN.[1:0] = 00 2MHz // 01 4MHz 10 8MHz 11 16MHz

char   wert_sec_10  = 0x00;     // 1/10 sec
char   wert_sec_1   = 0x00;     // 1 sec
char   wert_sec_6   = 0x00;     // 60 sec
char   wert_min_1   = 0x00;     // 1 -9 min

char   SSeg[10] = {   1+2+4+8+16+32,        //0
                      2+4,                  //1
                      1+4+8+32+64,          //2
                      1+2+4+8+64,           //3
                      2+4+16+64,            //4
                      1+2+8+16+64,          //5
                      1+2+8+16+32+64+128,   //6
                      2+4+8,                //7
                      255 -128,             //8
                      1+2+4+8+16+64         //9
                            };
```

Unter `// Function PROTOTYPES` wird die Art des Rückgabewertes festgelegt. Also ist der Rückgabewert ein „Charakter" oder ein „Integer". Im Falle „void" bedeutet es, dass Parameter vermieden werden. Dieser Teil ändert sich nicht zum Code von „Blinky-stripped". Der entsprechende Programmcode stellt sich wie folgt dar:

```
void PORT_Init (void);
void Timer3_Init (int counts);
void Timer3_ISR (void);            //void OSCILLATOR_Init (void);
```

Die Definition der SysClock Daten findet sich im Programmcode unter `// MAIN Routine`. Auf Basis des oben erklärten Verhaltes der „SysClock Funktion" wurde „Blinky-stripped" im Wesentlichen so modifiziert dass, die SysClock alle 12 Takte (6μs) einen Integer-Wert (PORT_Init) in den „Counter" „wirft". Das heißt, dass alle 6μs der Counter um 1 hoch gezählt wird. Der Timer3_Reloadvalue (Reloadzeit: $T_{rel} = 0,4s$) macht den Counter dann einmal komplett leer. Und der „Interrupt" (10Hz) gibt ein Bit-Signal aus. Dies wird mit einer While-Schleife realisiert.

```
void main (void) {

    // disable watchdog timer
    WDTCN = 0xde;
```

```
        WDTCN = 0xad;
        // set internal oscillator frequency
        OSCICN &= 0xfc;
        OSCICN |= oscmask;
        //OSCILLATOR_Init ();
        PORT_Init ();
        Timer3_Init ( SYSCLK / 12 / 10);        // Init Timer3 to generate interrupts
                                                // at a 10Hz rate. for 2 MHz
        EA = 1;                                 // enable global interrupts
        while (1) {                             // spin forever
        }
}
```

Asl nächstes muss die Definition von `// Timer3_Init` vorgenommen werden. Diese weist keine Unterschiede zu „Blinky-stripped" auf. Daher soll dieser lediglich dargestellt und nicht genauer beschrieben werden. Die Kommentare neben dem Programmcode sollen für die Beschreibung ausreichen.

```
void Timer3_Init (int counts)
{
    TMR3CN = 0x00;                          // Stop Timer3; Clear TF3;
                                            // use SYSCLK/12 as timebase
    TMR3RL  = 0xffff-counts;                // Init reload values
    TMR3    = 0xffff;                       // set to reload immediately
    EIE2   |= 0x01;                         // enable Timer3 interrupts
    TMR3CN |= 0x04;                         // start Timer3
}
```

Die Port- und Registerdefinition mit `// PORT_Init` wurde lediglich um die Ports 0 und 2 erweitert. Zudem wird definiert, wie die 8-Bitwerte an den Ports ausgegeben werden. Um die Eigenschaften der Ports zu bestimmen, wurde das Datenblatt (User Guide Seite 173 Figure: 17.14 „Port1 Output Mode Register") zu Hilfe genommen. Die folgende Abbildung zeigt diese Tabelle.

Figure 17.14. P1MDOUT: Port1 Output Mode Register

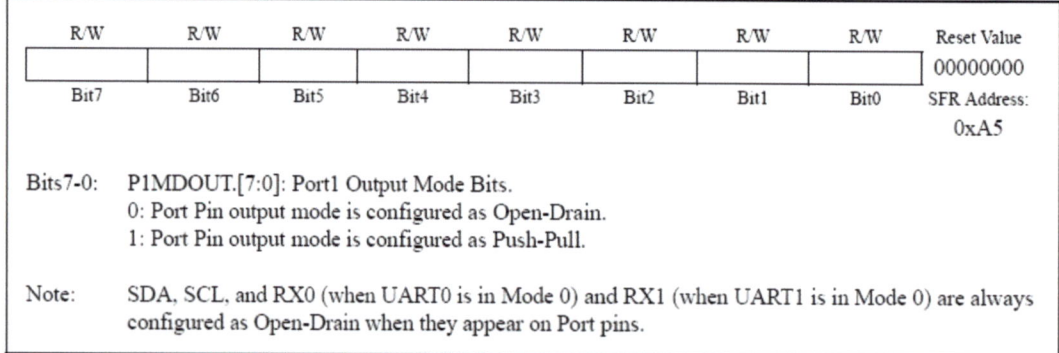

Abbildung 8: Eigenschaften von Port 0-7 aus dem Datenblatt.

In ihr heißt es z.B. für Port 1: "Output Mode Bits: 0=Open-Drain und 1= Push-Pull". Die PxMDOUT-Register sind die Modusschalter für die einzelnen Aus- und Eingabebits eines Ports. Das bedeutet bei 0 wird mit einer externen Spannungsquelle gearbeitet und bei 1 die interne Spannung von 3,3V verwendet. Da für unsere 7-Segmentanzeige 3,3V ausreichend sind, werden die benötigten Ports (0, 1, 2) auf „1" gesetzt. Diese „1" wird wie folgt eingegeben. Das Register hat 8 Bits und ist eine binär dargestellte Zahl. Wenn alle 8 Bits auf 1 gesetzt sein sollen, muß man binär 1111 1111 eingeben. Dezimal heißt das: 255 und Hexadezimal: FF (15*16 + 1*16). Im C-Code wird das mit „0xff" realisiert. Somit ergibt der folgende Programmteil:

```
void PORT_Init (void)
{
    XBR2    = 0x40;
    P0MDOUT = 0xff;
    P1MDOUT = 0xff;
    P2MDOUT = 0xff;
```

Der wichtigste Teil des Code, ist der „Timer 3_Interrupt Service Routine (ISR)"; er findet sich unter: // Timer3_ISR und soll im Folgenden etwas genauer beleuchtet werden. Im Keil Compiler gibt es eine Interrupt Service Routine, welche bei jedem Interrupt ausgeführt wird. In dieser Routine werden die benötigten Bits umgesetzt. In unserem Code wird die Routine also zehnmal/Sekunde ausgeführt. Das heißt, unsere Uhr läuft im konstanten Takt (Timer 0) auf 10tel-Sekunden Basis und zählt auf 8-Bit-Ebene hoch. Der Timer 1 schaltet synchron und gibt das Ausgangsignal.

```
void Timer3_ISR (void) interrupt 14
{
    TMR3CN &= ~(0x80);
```

Der Befehl „void" gibt lediglich den Rückgabewert an. Da dieser sich im Vergleich zu „Blinky-stripped" nicht verändert hat, soll auf ihn nicht weiter eingegangen werden.

```
P0 = SSeg[wert_sec_1];
P1 = SSeg[wert_sec_6];
P2 = SSeg[wert_min_1];
```

Hiermit wurde definiert, dass an Port 0 die Sekunden, an Port 1 die Zehner-Sekunden-Stelle und an Port 2 die Minuten ausgegeben werden. Zusätzlich wurde bestimmt, dass die Zehntelsekunden wert_sec_10 nicht ausgegeben werden.

```
wert_sec_10 = wert_sec_10 +1;      // 10telSekunden um 1 hochzählen
  if(wert_sec_10== 0x0a)           // Wenn 10tel Sekunden 10(hex=a) erreicht hat
  {
   wert_sec_10 = 0x00;             // 10tel Sekunden zu 0 initialisieren
   wert_sec_1 = wert_sec_1 + 1;    // Dann Sekunden um 1 hochzählen
     if(wert_sec_1 ==0x0a)         // Wenn Sekunden 10 (in hex=a) erreicht hat
     {
       wert_sec_1 = 0x00;          // Sekunden zu initialisieren
       wert_sec_6 = wert_sec_6+1;// 10erSekunden-Stelle um 1 hochzählen
       if(wert_sec_6 == 0x06)      // Wenn 10 Sekunden-Stelle 6 (in hex=6) erreicht hat
       {
         wert_sec_6 = 0x00;        // Zehner Sekunden Stelle zu 0 initialisieren
           wert_min_1 = wert_min_1+1;
           if(wert_min_1 == 0x0a) wert_min_1 = 0x00;
```

Dieser Teil beschreibt, dass pro Takt um eine Zehntelsekunde hochgezählt wird. Wenn der Wert 10 (hex=a) erreicht ist, soll Port 0 ein Bitmuster „weitergehen", also an Port 0 z.B. von Bitmuster 5 auf Bitmuster 6 umschalten. Wenn die Sekunden den Wert 10 (hex=a) erreicht haben, soll er wieder auf das Bitmuster für 0 „Springen". Weiter wird beschrieben, dass die Sekunden um 1 hochgezählt werden usw. Um die Beschreibung übersichtlich zu gestalten, wurde ein Kommentar neben die entsprechenden Programmbefehle gesetzt. Mit der Weiterführung dieses Algorithmus ist es möglich, eine vollständige 24-Stunden-Uhr zu entwickeln. Darüber hinaus besteht die Möglichkeit, den Programmcode zu einem Wecker zu erweitern und dabei ein Signal auszugeben, welches zum Beispiel eine Klingel betätigt.

4 Zusammenfassung und Ausblick

Schon zu Beginn der praktischen Versuche stellte sich heraus, dass die mitgelieferte IDE Entwicklungsumgebung für die C-Programmierung ausreichend ist und somit „Microsoft Visual C++ 2010 Express" gar nicht benötigt wird. Sie diente lediglich dazu, sich mit den Grundlagen der Programmierung vertraut zu machen. Allerdings war die Benutzung von Visual C++ 2010 dahingehend sinnvoll, weil das Programm die verschiedenen „Typen" im Programmcode farblich abgrenzt. Das heißt, Kommentare werden in grün dargestellt, Funktionen in blau und Fehler in rot. Dies macht den Programmcode sehr übersichtlich. Diesen Service bietet die IDE von Silicon Laboratories nicht.

Es wurde die Baugruppe mit den zwei 7-Segmentanzeigen nach der Beschreibung unter Abschnitt 3.3.1. aufgebaut. Diese wurden anschließend über Port 0 und 1 mit dem Mikrocontroller verbunden. Im nächsten Schritt wurde der Programmcode aus Anhang 4 mit der IDE auf den 8051 übertragen und ausgeführt. Daraufhin begannen die beiden Anzeigen im Sekundentakt die Zahlen von 0-19 auszugeben. Anschließend wurde eine weitere Anzeige an Port 2 angeschlossen, welche sofort die Minuten von 1 bis 9 anzeigte. Nachdem das Entwicklungsboard vom Rechner getrennt wurde, lief die Digitaluhr weiter. Die nächste Abbildung zeigt die 16te Sekunde auf der entwickelten Digitaluhr. In der nebenstehenden Tabelle sind die nötigen Ressourcen für die beschriebene Uhr (bis 10 Min.) aufgeführt. Damit wurde, obwohl die Uhr nur bis 10 Min. anzeigt, die Zielsetzung dieses Projektes erreicht.

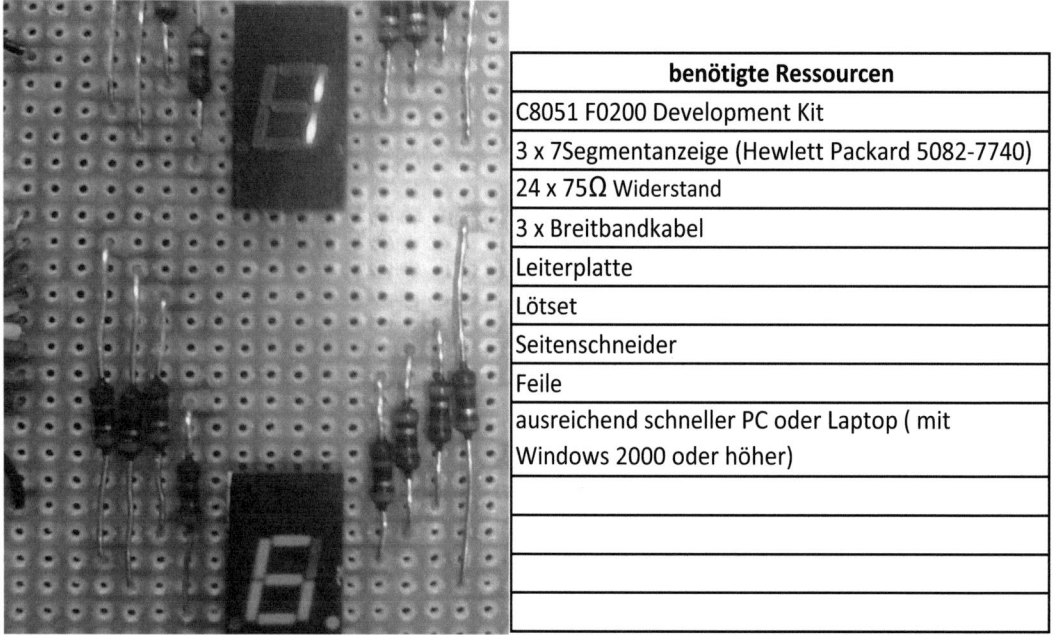

benötigte Ressourcen
C8051 F0200 Development Kit
3 x 7Segmentanzeige (Hewlett Packard 5082-7740)
24 x 75Ω Widerstand
3 x Breitbandkabel
Leiterplatte
Lötset
Seitenschneider
Feile
ausreichend schneller PC oder Laptop (mit Windows 2000 oder höher)

Abbildung 9: Sekunde 16 der laufenden Uhr und die benötigten Ressourcen des Projektes

Um die Digitaluhr optisch übersichtlicher zu gestalten, empfiehlt es sich, die Anzeigen nicht übereinander, sondern dicht nebeneinander zu platzieren und die benötigten Widerstände hinter der Leiterplatte zu verbergen. Wie bereits weiter oben erwähnt, besitzt der Mikrocontroller acht Parallelports (0-7). Damit wäre es möglich, durch Erweiterung des Programmcodes eine komplette Digitaluhr (22:17:19) zu entwickeln. Dieses Vorgehen wäre jedoch sehr impraktikabel und aufwendig, weil der Programmcode und die Hardware viel zu viele Ressourcen in Anspruch nehmen würden.

Literaturverzeichnis

[ITG1, 2010] Busse, Jürgen. 2010. *Informationstechnologie Teil 1: Grundlagen.* Hamburg, 2010.

[MFE, 2009] Hofmann, Michael. 2009. *Mikrocontroller für Einsteiger.* Poing : Franzis Verlag, 2009.

[ITH, 2005] Kreuzburg, Armin. 2005. *IT Handbuch.* Braunschweig : Westermann Schulbuchverlag, 2005.

[8051] Rolf, Klaus. 1999. *Die Mikrokontroller 8051, 8052 und 80C517.* Zürich : vdf Hochschulverlag AG an der ETH zürich, 1999.

[MT, 2009] Wüst, Klaus. 2009. *Mikroprozessortechnik.* Wiesbaden : Vieweg+Teubner GWV Fachverlage, 2009.

Internetquellen

[URL 1]: http://microcontroller.com/Microcontrollers/

[URL 2]: http://www.silabs.com/products/mcu/Pages/8051-microcontroller.aspx

[URL 3]: http://www.silabs.com/Support%20Documents/Software/8051_Instruction_Set.pdf

Anhang

Anhang 1: C8051 Interrupts_Online_FINAL

Anhang 2: Datenblatt C8051F02x Development Kit

Anhang 3: Programmcode „Blinky stripped"

Anhang 4: Programmcode „Uhr"

In this lecture we will look at the interrupt organization of C8051F900, how interrupts may be enabled or disabled and how to prioritize the interrupts. We will look at the polling sequence and learn about the interrupt pending flags for various interrupt sources. Also, we will discuss the external interrupts available in C8051F900.

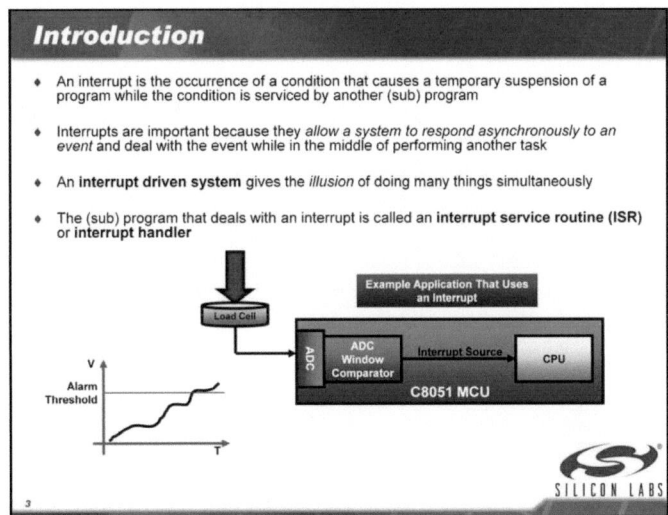

The picture above shows a load cell which is a transducer that provides an analog output voltage proportional to the force applied to the cell. Suppose this load cell is measuring the force applied to a threaded stem through a drive train and we don't want to exceed a certain rating. The MCU pictured may be controlling the motor driving the drive train and we want to stop the motor if the force applied exceeds our threshold shown in the graph. The internal Analog to Digital Converter (ADC) uses a windowing function whereby we can set the threshold limits and if the converted value is outside that window we can set an alarm. That alarm in this system is our interrupt and we can act on that to shut down the motor. The same can be applied to motor current for example. There are many different sources for interrupts both external (as we have illustrated) and internal to the MCU (things like timers and communications peripherals) and every application has a different set of requirements that drive the system/firmware architecture.

What is an interrupt? An interrupt is the occurrence of a condition that causes a temporary suspension of a program while the condition is serviced by another (sub) program. In our example it was excessive force on the load cell that triggered the ADC alarm. Interrupts change the program flow and allows the firmware to operate asynchronously to events that occur in the system. If the firmware was doing some trivial task when the alarm in our example occurred, we would vector immediately (within latency) to the critical task of shutting down the motor without having to wait for that other task to complete. When we are done with the critical function the firmware would continue what it was doing. Every application uses a different Interrupt mechanism allowing a system to respond asynchronously to an event and deal with the event while another is executing. The program that is executed when an interrupt occurs is called an **interrupt service routine (ISR)**. In this routine the firmware performs the required task like shutting down the motor. Other examples would be storing received data from a communications peripheral or changing the PWM duty cycle value of a timer to name a few.

Introduction

- ♦ The ISR executes in response to the interrupt and generally performs an input or output operation to a device

- ♦ When an interrupt occurs, the main program temporarily suspends execution and branches to the ISR

- ♦ The ISR executes, performs the desired operation, and terminates with a "return from interrupt" (RETI) instruction
  - ➢ The RETI instruction is different from the normal "RET" instruction

SILICON LABS

When an interrupt occurs, the main program temporarily suspends execution and branches to the ISR. The ISR executes, performs the operation, and terminates with a "return from interrupt" (RETI) instruction. We saw how this operation fits into a real example, however, a lot goes on at the CPU level to service an interrupt. It is important to note that there are two ways to change the execution flow in firmware. The first is as we described with the asynchronous events or interrupts. Another method is to use what is called a sub-routine. This is a firmware function located in the program memory that the firmware "calls" to execute a specific function. To return to the location in memory where the CPU was operating previously, a "Return" or "RET" instruction is used which restores the saved address. This is different with an interrupt return as they use the "Return from interrupt" or "RETI" instruction which also enables the interrupts.

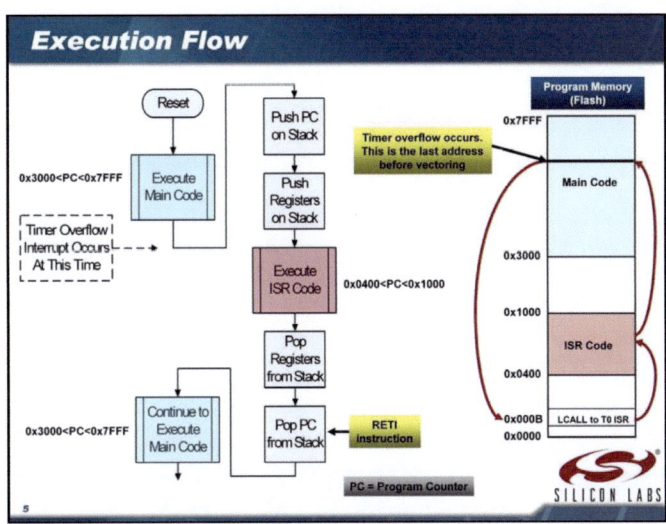

Here is a flow diagram of what happens when an interrupt occurs in the system, in this case a timer has overflowed (reached its maximum count value). Before we begin let's cover a couple of items we will need to be familiar with. The first is a stack which is an area of the data memory (RAM) that is allocated to store registers and variable contents. The act of pushing and popping (onto or off a stack) is used when referencing a stack and it refers to storing items into the memory and retrieving them out of the memory. Initially the firmware is running the main application code when the event occurs denoted by the dotted box. The program counter is within the main code section between 0x3000 and 0x7FFF. Once the event is detected the application firmware has to complete the current instruction before it can service the interrupt. Once the instruction completes the current value of the program counter (PC) is pushed onto the stack so we save where the main code was executing. Registers can be pushed onto the stack as needed. Once complete, the firmware can vector to the ISR. At this point we are executing code in the ISR Code (shaded red) and the value of the program counter has changed. The execution of the main program is effectively suspended while the ISR code is executed. Once the ISR completes, any registers pushed on the stack are popped off and then the PC is popped from the stack. Notice that the pushing onto the stack and the popping off of the stack happen in reverse order as the stack operates as a first in last out memory. By popping the registers and the PC we effectively restored that state of the CPU to the point before the event triggered the ISR and we can continue executing the main code from the next instruction.

Let's take a little deeper look at the interrupt vectoring mechanism. We touched a little on how the CPU vectors to the ISR which is a routine located in another region of the program memory. Once an interrupt is pending the interrupt controller will provide the address of the ISR vector location. The Interrupt vector listed is used to store an instruction that causes the code to branch execution to the location of the ISR in memory. Let's take a look at Timer 0 Overflow as an example. When the timer overflows the interrupt controller will pass the address 0x000B to the PC of the CPU. At the location 0x000B a jump instruction is found to the location in memory that is the first instruction of the ISR.

Associated with any ISR is something called latency which is the time it takes from the interrupt pending event to when the first instruction in the ISR executes. From our discussion of the interrupt process you can tell that there is latency from completing the current instruction and vectoring the ISR itself. In Silicon Labs MCUs, the latency is as fast as 7 system clock cycles.

♦ In the event of two or more simultaneous interrupts or an interrupt occurring while another is being serviced, there is both a <u>fixed priority **order**</u> and <u>two programmable priority **levels**</u> to schedule the interrupts

♦ If two interrupts are recognized simultaneously, the interrupt with the higher priority **level** is serviced first

♦ If both interrupts have the same priority **level**, the fixed priority **order** determines which is serviced first

6

What happens when two or more interrupts occur simultaneously or an interrupt occurring while another is being serviced? To schedule the interrupts there is both a *fixed priority order* and a *two-level priority scheme*. The fixed priority order is pre-determined, but the interrupt priority level is programmable. We will take a look at that in the upcoming slides.

Programmable Interrupt Priority Levels

♦ Each interrupt source can be individually programmed to one of **two priority levels**, low or high, through an associated interrupt priority bit in the SFRs **IP**, **EIP1** and **EIP2**

♦ These three SFRs are cleared after a system reset to place all interrupts at low priority by default

♦ The two priority *levels* allow an ISR to be interrupted by an interrupt of higher priority than the current one being serviced

♦ A low priority ISR is pre-empted by a high priority interrupt
  ➢ A high priority interrupt cannot be pre-empted

7

There are three SFRs - **IP**, **EIP1** and **EIP2** – which can be used to individually program the priority levels (low or high) of each interrupt source. These three SFRs are cleared after a system reset so that all the interrupts are placed at low priority by default. The two priority levels allow an ISR to be interrupted by an interrupt of higher priority than the current one being serviced. A low priority ISR is pre-empted by a high priority interrupt. A high priority interrupt cannot be pre-empted. Having two priority levels is useful because some events require immediate action, while some other events can tolerate some delay in the response time

Fixed Priority Order

Interrupt Source	Interrupt Vector	Priority Order
Reset	0000	Top
External Interrupt 0 (/INT0)	0003	0
Timer 0 Overflow	000B	1
External Interrupt 1 (/INT1)	0013	2
Timer 1 Overflow	001B	3
UART0	0023	4
Timer 2 Overflow	002B	5
SPI 0	0033	6
SMBus Interface	003B	7
smaRTClock Alarm	0043	8
ADC0 Window Comparator	004B	9

- Each interrupt source has a defined order that they would be serviced if pended simultaneously.
  - Example: UART0 gets serviced before SPI0 if the interrupts are posted at the same time

- Reset has the highest priority followed by the External Interrupt 0 and so on…

SILICON LABS

The priority order is a fixed value determined when the IC was designed. The priority order is used in order to determine which interrupt would get serviced first if two interrupts of the same programmable level were pended at the same time. The lower the priority number the higher the priority when getting serviced. For example, Timer 0 Overflow (priority order 1) would get serviced before the Timer 1 Overflow (priority order 3) if both were pended at the same time because Timer 0 has the lower priority order number. This priority number is used in conjunction with the priority registers.

Priority Level Example

- Critical tasks receive programmable priority level set to HIGH
  - Motor over current shut down
  - Analog transducer monitoring
  - Safety critical systems

- Non-critical tasks receive programmable priority level set to LOW
  - User input
  - Display units

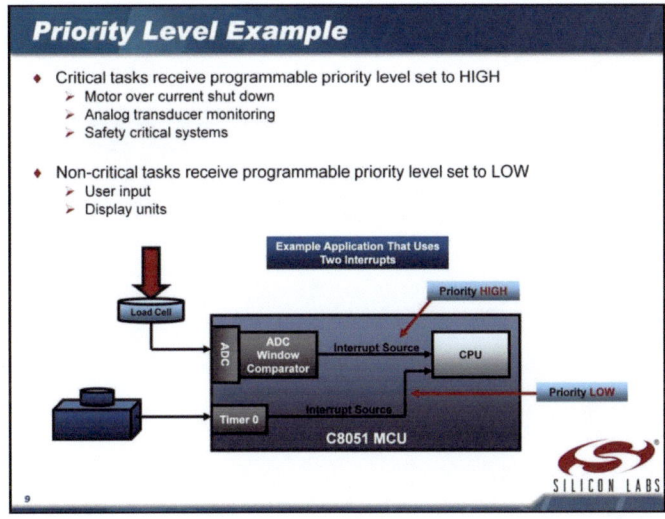

SILICON LABS

Again, the C8051F family of MCUs have two priority levels, high and low. Each interrupt source can be set to one of the two priority levels. It is within these levels that we then use the priority order if two interrupts of the same level are pended. Let's look again at our example. Let's say that the ADC0 Window Comparator is used for our load cell source and Timer 0 is used to time our user interface scanning like those for buttons. The ADC needs to be serviced with minimal latency (to save the motor) compared to our user buttons as any delay in those really won't affect operation. If we left the settings at their default values the timer ISR would execute before the ADC since its fixed priority is higher. We can set the ADC priority to high and the timer 0 to low and in order to ensure that if they are both pended at the same time then the ADC gets serviced first. Notice that the priority level control overrides the priority order.

Interrupt pending flags are the inputs to the peripheral's interrupt logic that generates the interrupt to the CPU. These flags are found within the SFRs of the specific peripheral. Some interrupt pending flags are automatically cleared by the hardware when the CPU vectors to the ISR. **However, most are not cleared by the hardware and must be cleared by software before returning from the ISR.** If an interrupt pending flag remains set after the CPU completes the RETI instruction, a new interrupt request will be generated immediately and the CPU will re-enter the ISR after the completion of the next instruction. When the interrupt flag is set to a '1', the interrupt will be posted, regardless of whether it is set by hardware or software. Any unused hardware interrupt can be "overtaken" and used as a software interrupt by explicitly setting the interrupt flag. This feature can also be used to "test" ISRs.

In order for an interrupt to be serviced by the CPU it needs to be enabled. There are three SFRs- IE, EIE1 and EIE2 – which are used to individually enable or disable each of the interrupt sources. There is also a global enable/disable bit, EA (IE.7) that is cleared to disable all interrupts or set to turn on all interrupts. Thus to enable any interrupt, two bits must be set - the corresponding individual enable bit and the global enable bit. If interrupts are disabled, the interrupt pending flags are ignored by the hardware and program execution continues as normal. There are some exceptions to the interrupt structure. The Programmable Counter Array (PCA) requires each of the capture compare interrupt sources to be enabled in addition to the PCA interrupt source. Contrast this to the typical scenario where there is one interrupt source with either one or more interrupt pending flags associated with it and the single interrupt enable for the source.

X

- Pins for the two external interrupt sources (/INT0 and /INT1) are allocated and assigned by the crossbar

- They are configured by bits IT0 (TCON.0) and IT1 (TCON.2)

- IE0 (TCON.1) and IE1 (TCON.3) serve as the interrupt-pending flags

- They are enabled using bits EX0 (IE.0) and EX1 (IE.1)

- The external interrupt sources can be programmed to be level-activated (low) or transition-activated (falling edge) on /INT0 or /INT1
  - If a /INT0 or /INT1 external interrupt is configured as edge-sensitive, the corresponding interrupt-pending flag is automatically cleared by hardware when the CPU vectors to the ISR
  - When configured as level sensitive, the interrupt-pending flag follows the state of the external interrupt's input pin
    - The external interrupt source must hold the input active until the interrupt request is recognized
    - It must then deactivate the interrupt request before execution of the ISR completes, or else another interrupt request will be generated

SILICON LABS

12

The two external interrupt sources (/INT0 and /INT1) do not have dedicated pins on the microcontroller, they are allocated and assigned by the digital crossbar by configuring bits IT0 (TCON.0) and IT1 (TCON.2). These external interrupt sources can be programmed to be level-activated (low) or transition-activated (falling edge).

Interrupt Organization

- The C8051F9xx family supports 19 interrupt sources, including:
  - 2 external interrupts (/INT0, /INT1)
  - 4 timer interrupts (Timer 0 through 3 Overflow)
  - 1 serial port interrupts (UART0)

- Each interrupt source has one or more associated interrupt-pending flag(s) located in an SFR

- When a peripheral or external source meets a valid interrupt condition, the associated interrupt-pending flag is set to 1
  - These interrupt flags are "level sensitive" in that if the flag is not cleared in the ISR by either hardware or software, the interrupt will trigger again, even if the event that originally caused the interrupt did not occur again

- All interrupts are disabled after a system reset and enabled individually by software

SILICON LABS

13

Now that we understand what an interrupt does, let's take a look at an MCUs implementation of interrupts. The Silicon Labs C8051F93x-92x family supports 19 interrupt sources meaning that the CPU can vector to 19 different locations based on the interrupting event. The 19 sources are comprised of internally generated events as well as the ability to provide an external trigger to interrupt the CPU. A peripheral generates an event by setting a bit in a Special Function Register (SFR) that signals the interrupt controller to interrupt the CPU. Each of the 19 interrupt sources has one or more associated interrupt-pending flag(s) located in an SFR so we can see that there are actually more sources overall. One example of multiple flags is the UART peripheral where the receive and transmit interrupts source a single vector. In general, when a peripheral or an external source meets a valid interrupt condition, the associated interrupt-pending flag is set to 1. When an interrupt is pending (flag bit in SFR is set) the system will service the interrupt. Within the ISR, the firmware should clear the pended bit if the hardware doesn't clear it automatically. Care must be taken to verify which interrupts get cleared automatically and which ones don't because any pended bit will trigger an ISR whether it just completed or not. If the pended bit is not cleared the CPU will return from the ISR and then vector right back to the same one. In order to start using any interrupts each one has to be enabled as they are disabled by default after a system reset.

♦ Interrupt vector table (Priority 0 – 9)

Interrupt Source	Interrupt Vector	Priority Order	Pending Flag	Enable Flag	Priority Control
Reset	0000	Top	None	Always Enabled	Always Highest
External Interrupt 0 (/INT0)	0003	0	IE0 (TCON.1)	EX0 (IE.0)	PX0 (IP.0)
Timer 0 Overflow	000B	1	TF0 (TCON.5)	ET0 (IE.1)	PT0 (IP.1)
External Interrupt 1 (/INT1)	0013	2	IE1 (TCON.3)	EX1 (IE.2)	PX1 (IP.2)
Timer 1 Overflow	001B	3	TF1 (TCON.7)	ET1 (IE.3)	PT1 (IP.3)
UART0	0023	4	RI0 (SCON0.0) TI0 (SCON0.1)	ES0 (IE.4)	PS0 (IP.4)
Timer 2 Overflow	002B	5	TF2H (TMR2CN.7) TF2L (TMR2CN.6)	ET2 (IE.5)	PT2 (IP.5)
SPI0	0033	6	SPIF (SPI0CN.7) WCOL (SPI0CN.6) MODF (DPI0CN.5) RXOVRN (SPI0CN.4)	ESPI0 (IE.6)	PSPI0 (IP.6)
SMBus Interface	003B	7	SI (SMB0CN.0)	ESMB0 (EIE1.0)	PSMB0 (EIP1.0)
smaRTClock Alarm	0043	8	ALRM (RTC0CN.2)	EARTC0 (EIE1.1)	PARTC0 (EIP1.1)
ADC0 Window Comparator	004B	9	AD0WINT (ADC0CN.3)	EWADC0 (EIE1.2)	PWADC0 (EIP1.2)

14

This table on this slide and the next shows the list of all the interrupt sources in the F9xx family as an example. The associated interrupt pending flag is also shown. For example, when the Timer 0 overflows, the corresponding interrupt pending flag TF0 (which is bit 5 in TCON SFR) will be set. You can see from the table that there are several peripherals that have multiple pending flags associated with the single interrupt source. The SPI0 module has 4 sources for the interrupt. That means that the software will have to determine the interrupting source in the ISR. An interrupt will be generated for any of the flags if it has been enabled. In order to enable an interrupt the associated interrupt enable bit must be set. There are three registers that contain all of the interrupt enables for the associated peripherals, IE, EIE1 and EIE2. For Timer 0 Overflow to generate an interrupt, the enable flag ET0 (which is bit 1 of IE SFR) must be set.

♦ Interrupt vector table (priority 10-18)

Interrupt Source	Interrupt Vector	Priority Order	Pending Flag	Enable Flag	Priority Control
ADC0 End of Conversion	0530	10	AD0INT (ADC0STA.5)	EADC0 (EIE1.3)	PADC0 (EIP1.3)
Programmable Counter Array	005B	11	CF (PCA0CN.7) CCFn (PCA0CN.n)	EPCA0 (EIE1.4)	PPCA0 (EIP1.4)
Comparator 0	0063	12	CP0FIF (CPT0CN.4) CP0RIF (CPT0CN.5)	ECP0 (EIE1.5)	PCP0 (EIP1.5)
Comparator 1	006B	13	CP1FIF (CPT1CN.4) CP1RIF (CPT1CN.5)	ECP1 (EIE1.6)	PCP1 (EIP1.6)
Timer 3 Overflow	0073	14	TF3H (TMR3CN.7) TF3L (TMR3CN.6)	ET3 (EIE1.7)	PT3 (EIP1.7)
VDD/DC+ Supply Monitor Early Warning	007B	15	VDDOK(VDM0CN.5)	EWARN (EIE2.0)	PWARN (EIP2.0)
Port Match	0083	16	None	EMAT (EIE2.1)	PMAT (EIP2.1)
smaRTClock Oscillator Fail	008B	17	OSCFAIL (RTC0CN.5)	ERTC0F (EIE2.2)	PFRTC0F (EIP2.2)
SPI 1	0093	18	SPIF(SPI1CN.7) WCOL(SPI1CN.6) MODF(SPI1CN.5) RXOVRN(SPI1CN.4)	ESPI1 (EIE2.3)	PSPI1 (EIP2.3)

15

Here are some more interrupt sources. Notice also the Interrupt Vector addresses. Previously we discussed how the interrupt logic loads the PC with the Interrupt Vector number and provides a mechanism to jump to the ISR routine in flash. The address of the ISR is part of the instruction found at the Interrupt Vector location.

TCON Register Description

- SFR register address: 0x88
- SFR Page: 0

Bit	Symbol	Description
7	TF1	**Timer 1 Overflow Flag** Set by hardware when Timer 1 overflows. This flag can be cleared by software but is automatically cleared when the CPU vectors to the Timer 1 interrupt service routine (ISR). 0: No Timer 1 overflow detected 1: Timer 1 has overflowed
6	TR1	**Timer 1 Run Control** 0: Timer 1 disabled 1: Timer 1 enabled
5	TF0	**Timer 0 Overflow Flag** Same as TF1 but applies to Timer 0 instead. 0: No Timer 0 overflow detected 1: Timer 0 has overflowed
4	TR0	**Timer 0 Run Control** 0: Timer 0 disabled 1: Timer 0 enabled
3	IE1	**External Interrupt 1** This flag is set by hardware when an edge/level of type defined by IT1 is detected. It can be cleared by software but is automatically cleared when the CPU vectors to the External Interrupt 1 ISR if IT1=1. This flag is the inverse of the /INT1 input signal's logic level when IT1=0.
2	IT1	**Interrupt 1 Type Select** 0: /INT1 is level triggered 1: /INT1 is edge triggered
1	IE0	**External Interrupt 0** Same as IE1 but applies to IT0 instead.
0	IT0	**Interrupt 0 Type Select** 0: /INT0 is level triggered 1: /INT0 is edge triggered

16

Why do we show the Timer Control Register when talking about interrupts? It is kind of strange, but external interrupts /INT0 and /INT1 are controlled by the TCON register. This is still maintained even in new 8051 derivatives to preserve backward compatibility with the original 8051 architecture. This register controls the sensitivity of the interrupt input (edge sensitive or level sensitive) as well as providing the interrupt pending flags.

Interrupt Enable (IE) Register Description

- SFR register address: 0xA8
- SFR page: All pages

Bit	Symbol	Description
7	EA	**Enable All Interrupts** 0: Disable all interrupt sources 1: Enable each interrupt according to its individual mask setting
6	ESPI0	**Enable Serial Peripheral Interface (SPI0) Interrupt** This bit sets the masking of the SPI0 interrupts 0: Disable all DPI0 interrupts 1: Enable interrupt requests generated by SPI0
5	ET2	**Enable Timer 2 Interrupt** 0: Disable Timer 2 Interrupt 1: Enable interrupt requests generated by TF2I and TF2H flags
4	ES0	**Enable UART0 Interrupt** 0: Disable UART0 Interrupt 1: Enable UART0 Interrupt
3	ET1	**Enable Timer 1 Interrupt** 0: Disable Timer 1 Interrupt 1: Enable interrupt requests generated by TF1 (TCON.7)
2	EX1	**Enable External Interrupt 1** 0: Disable external interrupt 1 1: Enable interrupt request generated by the /INT1 pin
1	ET0	**Enable Timer 0 Interrupt** 0: Disable Timer 0 Interrupt 1: Enable interrupt requests generated by TF0
0	EX0	**Enable External Interrupt 0** 0: Disable external interrupt 0 1: Enable interrupt request generated by the /INT0 pin

17

The IE register is one of the three SFRs that are used to enable/disable interrupt sources. The other two SFRs are EIE1 and EIE2. The EA bit is used to globally enable/disable interrupts. It is a good idea to globally disable interrupts before all the initialization routines (for system clock, port I/O, timers, UARTs, ADC/DAC etc) are executed and then enable the interrupts at the end. This way spurious interrupts won't get generated while configuring the peripherals.

Extended Interrupt Enable 1 (EIE1)

- SFR register address: 0xE6

- SFR page: All pages

Bit	Symbol	Description
7	ET3	**Enable Timer 3 Interrupt** 0: Disable Timer 3 Interrupt 1: Enable interrupt requests generated by TF3L and TF3H flags
6	ECP1	**Enable Comparator1 Interrupt** 0: Disable CP1 interrupt 1: Enable interrupt requests generated by CP1RIF or CP1FIF flags
5	ECP0	**Enable Comparator0 Interrupt** 0: Disable CP0 interrupt 1: Enable interrupt requests generated by CP0RIF or CP0FIF flags
4	EPCA0	**Enable Programmable Counter Array (PCA0) Interrupt** 0: Disable all PCA0 interrupts 1: Enable interrupt requests generated by PCA0
3	EADC0	**Enable ADC0 End of Conversion Interrupt** 0: Disable ADC0 End of Conversion interrupt 1: Enable interrupt requests generated by the ADC0 End of Conversion Interrupt
2	EWADC0	**Enable Window Comparison ADC0 Interrupt** 0: Disable ADC0 Window Comparison Interrupt 1: Enable interrupt request generated by ADC0 Window Comparisons (AD0WINT)
1	ERTC0A	**Enable smaRTClock Alarm Interrupts.** 0: Disable smaRTClock Alarm interrupts 1: Enable interrupt requests generated by a smaRTClock Alarm
0	ESMB0	**Enable System Management Bus (SMBus0) Interrupt** 0: Disable all SMBus interrupts 1: Enable interrupt requests generated by SMB0

18

Here is another register with the interrupt enables for additional peripherals.

Extended Interrupt Enable 2 (EIE2)

- SFR register address: 0xE7

- SFR page: All pages

Bit	Symbol	Description
7:4	Unused	**Unused** Read = 0000b. Write = Don't care.
3	ESPI1	**Enable Serial Peripheral Interface (SPI1) Interrupt** 0: Disable all SPI1 interrupts 1: Enable interrupt requests generated by SPIF1
2	ERTC0F	**Enable smaRTClock Oscillator Fail Interrupt** 0: smaRTClock alarm interrupts 1: Enable interrupts generated by smaRTClock alarm
1	EMAT	**Enable Port Match Interrupt** 0: Disable port match interrupts 1: Enable interrupt request generated by a port match
0	EWARN	**Enable VDD/DC+ Supply Monitor Early Warning Interrupt** 0: Disable the VDD/DC+ Supply early warning interrupt 1: Enable interrupts generated by the VDD/DC+ suppy monitor

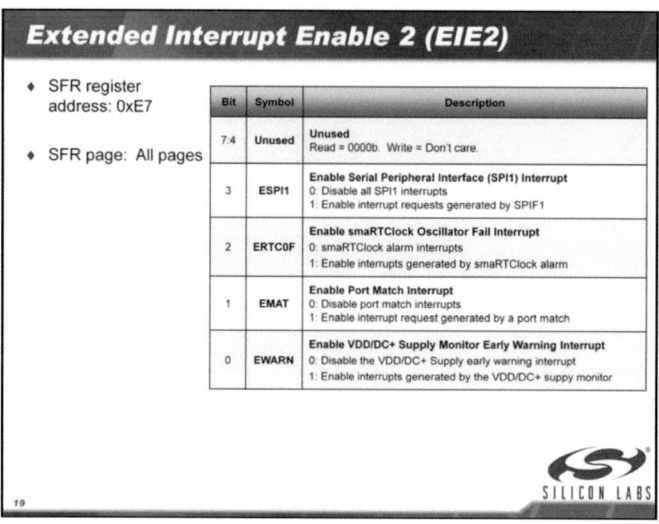

SILICON LABS

19

And some more enables. Remember there were three registers that contain all of the interrupt enables to the peripherals and devices within the different product families have different interrupt tables. Consult the datasheet for the specific device used to determine where the interrupt enables for each peripheral reside in the registers.

Interrupt Priority (IP)

- SFR register address: 0xB8

- SFR page: 0

Bit	Symbol	Description
7	-	UNUSED. Read=1, Write=don't care
	PSPI0	Serial Peripheral Interface (SPI0) Interrupt Priority Control 0: SPI0 interrupt set to low priority level 1: SPI0 interrupt set to high priority level
5	PT2	Timer 2 Interrupt Priority Control 0: Timer 2 Interrupt set to low priority level 1: Timer 2 interrupts set to high priority level
4	PS0	UART0 Interrupt Priority Control 0: UART0 Interrupts set to low priority level 1: UART0 interrupts set to high priority level
3	PT1	Timer 1 Interrupt Priority Control 0: Timer 1 Interrupt set to low priority level 1: Timer 1 interrupt set to high priority level
2	PX1	External Interrupt 1 Priority Control 0: External Interrupt 1 set to low priority level 1: External Interrupt 1 set to high priority level
1	PT0	Timer 0 Interrupt Priority Control 0: Timer 0 Interrupt set to low priority level 1: Timer 0 interrupt set to high priority level
0	PX0	External Interrupt 0 Priority Control 0: External Interrupt 0 set to low priority level 1: External Interrupt 0 set to high priority level

20

The IP register is one of the three SFRs that are used to set the priority level (default or high) of interrupt sources. The other two SFRs are EIP1 and EIP2. These registers follow the same format as the interrupt enable registers as the location of the priority bits follow those of the enables.

Extended Interrupt Priority 1 (EIP1)

- SFR register address: 0xF6

- SFR page: All pages

Bit	Symbol	Description
7	PT3	Timer 3 Interrupt Priority Control 0: Timer 3 interrupt set to low priority level 1: Timer 3 interrupt set to high priority level
6	PCP1	Comparator1 (CP1) Interrupt Priority Control 0: CP1 interrupt set to low priority level 1: CP1 interrupt set to high priority level
5	PCP0	Comparator0 (CP0) Interrupt Priority Control 0: CP0 interrupt set to low priority level 1: CP0 interrupt set to high priority level
4	PPCA0	Programmable Counter Array (PCA0) Interrupt Priority Control 0: PCA0 interrupt set to low priority level 1: PCA0 interrupt set to high priority level
3	PADC0	ADC0 End of Conversion Interrupt Priority Control 0: ADC0 End of Conversion interrupt set to low priority level 1: ADC0 End of Conversion interrupt set to high priority level
2	PWADC0	ADC0 Window Comparator Interrupt Priority Control 0: ADC0 Window interrupt set to low priority level 1: ADC0 Window interrupt set to high priority level
1	PRTC0A	smaRTClock Alarm Interrupt Priority Control 0: smaRTClock Alarm interrupt set to low level 1: smaRTClock Alarm interrupt set to high level
0	PSMB0	System Management Bus (SMBus0) Interrupt Priority Control 0: SMBus interrupt set to low priority level 1: SMBus interrupt set to high priority level

21

Here is the second interrupt priority register.

Extended Interrupt Priority 2 (EIP2)

- SFR register address: 0xF7

- SFR page: All pages

Bit	Symbol	Description
7:4	Unused	Unused Read = 0000b. Write = Don't care.
3	PSPI1	**Serial Peripheral Interface (SPI1) Interrupt Priority Control** 0: SPI1 interrupts set to low priority 1: SPI1 interrupts set to high priority
2	PRTC0F	**smaRTClock Oscillator Fail Interrupt Priority Control** 0: smaRTClock alarm interrupts set to low priority 1: smaRTClock alarm interrupts set to high priority
1	PMAT	**Port Match Interrupt Priority Control** 0: Port match interrupts set to low priority 1: Port match interrupts set to high priority
0	PWARN	**VDD/DC+ Supply Monitor Early Warning Interrupt Priority Control** 0: VDD/DC+ Supply early warning interrupt set to low priority 1: VDD/DC+ Supply early warning interrupt set to high priority

SILICON LABS

22

Here is the third interrupt priority register.

Learn More at the Education Resource Center

- Visit the Silicon Labs website to get more information on Silicon Labs products, technologies and tools

- The Education Resource Center training modules are designed to get designers up and running quickly on the peripherals and tools needed to get the design done
  - http://www.silabs.com/ERC
  - http://www.silabs.com/mcu

- To provide feedback on this or any other training go to:

  http://www.silabs.com/ERC and click the link for feedback

SILICON LABS

23

Visit the Silicon Labs Education Resource Center to learn more about the MCU products.

C8051F02x-DK

C8051F02x DEVELOPMENT KIT USER'S GUIDE

1. Kit Contents

The C8051F02x Development Kit contains the following items:

- C8051F020 Target Board
- C8051Fxxx Development Kit Quick-Start Guide
- Silicon Laboratories IDE and Product Information CD-ROM. CD content includes:
  - Silicon Laboratories Integrated Development Environment (IDE)
  - Keil 8051 Development Tools (macro assembler, linker, evaluation 'C' compiler)
  - Source code examples and register definition files
  - Documentation
  - C8051F02x Development Kit User's Guide (this document)
- AC to DC Power Adapter
- USB Debug Adapter (USB to Debug Interface)
- USB Cable

2. Hardware Setup using a USB Debug Adapter

The target board is connected to a PC running the Silicon Laboratories IDE via the USB Debug Adapter as shown in Figure 1.

1. Connect the USB Debug Adapter to the JTAG connector on the target board with the 10-pin ribbon cable.
2. Connect one end of the USB cable to the USB connector on the USB Debug Adapter.
3. Connect the other end of the USB cable to a USB Port on the PC.
4. Connect the ac/dc power adapter to power jack P1 on the target board.

Notes:
- Use the Reset button in the IDE to reset the target when connected using a USB Debug Adapter.
- Remove power from the target board and the USB Debug Adapter before connecting or disconnecting the ribbon cable from the target board. Connecting or disconnecting the cable when the devices have power can damage the device and/or the USB Debug Adapter.

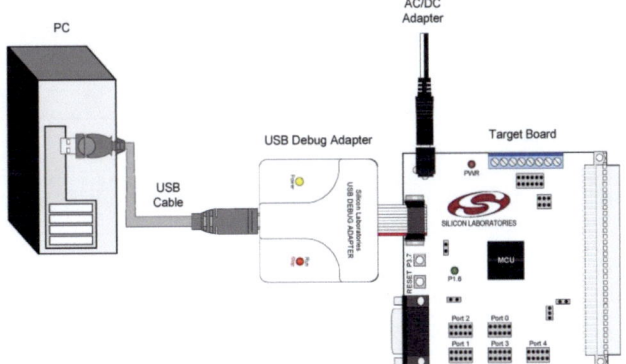

Figure 1. Hardware Setup using a USB Debug Adapter

Rev. 0.6 9/06   Copyright © 2006 by Silicon Laboratories   C8051F02x-DK

3. Software Setup

The included CD-ROM contains the Silicon Laboratories Integrated Development Environment (IDE), Keil software 8051 tools and additional documentation. Insert the CD-ROM into your PC's CD-ROM drive. An installer will automatically launch, allowing you to install the IDE software or read documentation by clicking buttons on the Installation Panel. If the installer does not automatically start when you insert the CD-ROM, run *autorun.exe* found in the root directory of the CD-ROM. Refer to the *readme.txt* file on the CD-ROM for the latest information regarding known IDE problems and restrictions.

4. Silicon Laboratories Integrated Development Environment

The Silicon Laboratories IDE integrates a source-code editor, source-level debugger and in-system Flash programmer. The use of third-party compilers and assemblers is also supported. This development kit includes the Keil Software A51 macro assembler, BL51 linker and evaluation version C51 'C' compiler. These tools can be used from within the Silicon Laboratories IDE.

4.1. System Requirements

The Silicon Laboratories IDE requirements:

- Pentium-class host PC running Microsoft Windows 98SE or later.
- One available COM or USB port.
- 64 MB RAM and 40 MB free HD space recommended.

4.2. Assembler and Linker

A full-version Keil A51 macro assembler and BL51 banking linker are included with the development kit and are installed during IDE installation. The complete assembler and linker reference manual can be found under the **Help** menu in the IDE or in the "*Silabs\MCU\hlp*" directory (A51.pdf).

4.3. Evaluation C51 'C' Compiler

An evaluation version of the Keil C51 'C' compiler is included with the development kit and is installed during IDE installation. The evaluation version of the C51 compiler is the same as the full professional version except code size is limited to 4 kB and the floating point library is not included. The C51 compiler reference manual can be found under the **Help** menu in the IDE or in the "*Silabs\MCU\hlp*" directory (C51.pdf).

4.4. Using the Keil Software 8051 Tools with the Silicon Laboratories IDE

To perform source-level debugging with the IDE, you must configure the Keil 8051 tools to generate an absolute object file in the OMF-51 format with object extensions and debug records enabled. You may build the OMF-51 absolute object file by calling the Keil 8051 tools at the command line (e.g. batch file or make file) or by using the project manager built into the IDE. The default configuration when using the Silicon Laboratories IDE project manager enables object extension and debug record generation. Refer to Applications Note **AN104 - Integrating Keil 8051 Tools Into the Silicon Laboratories IDE** in the "*SiLabs\MCU\Documentation\Appnotes*" directory on the CD-ROM for additional information on using the Keil 8051 tools with the Silicon Laboratories IDE.

To build an absolute object file using the Silicon Laboratories IDE project manager, you must first create a project. A project consists of a set of files, IDE configuration, debug views, and a target build configuration (list of files and tool configurations used as input to the assembler, compiler, and linker when building an output object file).

The following sections illustrate the steps necessary to manually create a project with one or more source files, build a program and download the program to the target in preparation for debugging. (The IDE will automatically create a single-file project using the currently open and active source file if you select **Build/Make Project** before a project is defined.)

4.4.1. Creating a New Project

1. Select **Project→New Project** to open a new project and reset all configuration settings to default.

2. Select **File→New File** to open an editor window. Create your source file(s) and save the file(s) with a recognized extension, such as .c, .h, or .asm, to enable color syntax highlighting.

3. Right-click on "New Project" in the **Project Window**. Select **Add files to project**. Select files in the file browser and click Open. Continue adding files until all project files have been added.

4. For each of the files in the **Project Window** that you want assembled, compiled and linked into the target build, right-click on the file name and select **Add file to build**. Each file will be assembled or compiled as appropriate (based on file extension) and linked into the build of the absolute object file.

   Note: If a project contains a large number of files, the "Group" feature of the IDE can be used to organize. Right-click on "New Project" in the **Project Window**. Select **Add Groups to project**. Add pre-defined groups or add customized groups. Right-click on the group name and choose **Add file to group**. Select files to be added. Continue adding files until all project files have been added.

4.4.2. Building and Downloading the Program for Debugging

1. Once all source files have been added to the target build, build the project by clicking on the **Build/Make Project** button in the toolbar or selecting **Project→Build/Make Project** from the menu.

   Note: After the project has been built the first time, the **Build/Make Project** command will only build the files that have been changed since the previous build. To rebuild all files and project dependencies, click on the **Rebuild All** button in the toolbar or select **Project→Rebuild All** from the menu.

2. Before connecting to the target device, several connection options may need to be set. Open the **Connection Options** window by selecting **Options→Connection Options...** in the IDE menu. First, select the appropriate adapter in the "Serial Adapter" section. Next, the correct "Debug Interface" must be selected. C8051F02x family devices use the JTAG debug interface. Once all the selections are made, click the OK button to close the window.

3. Click the **Connect** button in the toolbar or select **Debug→Connect** from the menu to connect to the device.

4. Download the project to the target by clicking the **Download Code** button in the toolbar.

   Note: To enable automatic downloading if the program build is successful select **Enable automatic connect/download after build** in the **Project→Target Build Configuration** dialog. If errors occur during the build process, the IDE will not attempt the download.

5. Save the project when finished with the debug session to preserve the current target build configuration, editor settings and the location of all open debug views. To save the project, select **Project->Save Project As...** from the menu. Create a new name for the project and click on **Save**.

SILICON LABORATORIES

Rev. 0.6

3

5. Example Source Code

Example source code and register definition files are provided in the "*SiLabs\MCU\Examples\C8051F02x*" directory during IDE installation. These files may be used as a template for code development. Example applications include a blinking LED example which configures the green LED on the target board to blink at a fixed rate.

5.1. Register Definition Files

Register definition files *C8051F020.inc and C8051F020.h* define all SFR registers and bit-addressable control/ status bits. They are installed into the "*SiLabs\MCU\Examples\C8051F02x*" directory during IDE installation. The register and bit names are identical to those used in the C8051F02x data sheet. Both register definition files are also installed in the default search path used by the Keil Software 8051 tools. Therefore, when using the Keil 8051 tools included with the development kit (A51, C51), it is not necessary to copy a register definition file to each project's file directory.

5.2. Blinking LED Example

The example source files *blink.asm* and *blinky.c* show examples of several basic C8051F02x functions. These include; disabling the watchdog timer (WDT), configuring the Port I/O crossbar, configuring a timer for an interrupt routine, initializing the system clock, and configuring a GPIO port. When compiled/assembled and linked this program flashes the green LED on the C8051F020 target board about five times a second using the interrupt handler with a C8051F020 timer.

6. Target Board

The C8051F02x Development Kit includes a target board with a C8051F020 device pre-installed for evaluation and preliminary software development. Numerous input/output (I/O) connections are provided to facilitate prototyping using the target board. Refer to Figure 2 for the locations of the various I/O connectors.

P1	Power connector (accepts input from 7 to 15 VDC unregulated power adapter)
J1	Connects SW2 to P3.7 pin
J3	Connects LED D3 to P1.6 pin
J4	JTAG connector for Debug Adapter interface
J5	DB-9 connector for UART0 RS232 interface
J6	Connector for UART0 TX (P0.0)
J8	Connector for UART0 RTS (P4.0)
J9	Connector for UART0 RX (P0.1)
J10	Connector for UART0 CTS (P4.1)
J11	Analog loopback connector
J12-J19	Port 0 - 7 connectors
J20	Analog I/O terminal block
J22	VREF connector
J23	VDD Monitor Disable
J24	96-pin Expansion I/O connector

Figure 2. C8051F020 Target Board

6.1. System Clock Sources

The C8051F020 device installed on the target board features a calibrated programmable internal oscillator which is enabled as the system clock source on reset. After reset, the internal oscillator operates at a frequency of 2 MHz (±2%) by default but may be configured by software to operate at other frequencies. Therefore, in many applications an external oscillator is not required. However, an external 22.1184 MHz crystal is installed on the target board for additional applications. Refer to the C8051F02x data sheet for more information on configuring the system clock source.

6.2. Switches and LEDs

Two switches are provided on the target board. Switch SW1 is connected to the RESET pin of the C8051F020. Pressing SW1 puts the device into its hardware-reset state. Switch SW2 is connected to the C8051F020's general purpose I/O (GPIO) pin through headers. Pressing SW2 generates a logic low signal on the port pin. Remove the shorting block from the header to disconnect SW2 from the port pins. The port pin signal is also routed to a pin on the J24 I/O connector. See Table 1 for the port pins and headers corresponding to each switch.

Two LEDs are also provided on the target board. The red LED labeled PWR is used to indicate a power connection to the target board. The green LED labeled with a port pin name is connected to the C8051F020's GPIO pin through headers. Remove the shorting block from the header to disconnect the LED from the port pin. The port pin signal is also routed to a pin on the J24 I/O connector. See Table 1 for the port pins and headers corresponding to each LED.

Table 1. Target Board I/O Descriptions

Description	I/O	Header
SW1	Reset	none
SW2	P3.7	J1
Green LED	P1.6	J3
Red LED	PWR	none

6.3. Target Board JTAG Interface (J4)

The JTAG connector (J4) provides access to the JTAG pins of the C8051F020. It is used to connect the Serial Adapter or the USB Debug Adapter to the target board for in-circuit debugging and Flash programming. Table 2 shows the JTAG pin definitions.

Table 2. JTAG Connector Pin Descriptions

Pin #	Description
1	+3 VD (+3.3 VDC)
2, 3, 9	GND (Ground)
4	TCK
5	TMS
6	TDO
7	TDI
8, 10	Not Connected

SILICON LABORATORIES

XXII

6.4. Serial Interface (J5)

A RS232 transceiver circuit and DB-9 (J5) connector are provided on the target board to facilitate serial connections to UART0 of the C8051F020. The TX, RX, RTS and CTS signals of UART0 may be connected to the DB-9 connector and transceiver by installing shorting blocks on headers J6, J8, J9 and J10.

J6 - Install shorting block to connect UART0 TX (P0.0) to the transceiver.
J9 - Install shorting block to connect UART0 RX (P0.1) to the transceiver.
J8 - Install shorting block to connect UART0 RTS (P4.0) to the transceiver.
J10 - Install shorting block to connect UART0 CTS (P4.1) to the transceiver.

6.5. Analog I/O (J11, J20)

Several C8051F020 analog signals are routed to the J20 terminal block and the J11 connector. Header J11 provides the ability to connect DAC0 and DAC1 outputs to several different analog inputs by installing a shorting block between a DAC output and an analog input on adjacent pins of J11. Refer to Table 3 for J20 terminal block connections and Table 4 for J11 pin definitions.

Table 3. J20 Terminal Block Pin Descriptions

Pin #	Description
1	CP0+
2	CP0-
3	DAC0
4	DAC1
5	AIN0.0
6	AIN0.1
7	VREF0
8	ADND (Analog Ground)

Table 4. J11 Connector Pin Descriptions

Pin #	Description
1	CP0+
2	CP0-
3	DAC0
4	DAC1
5	CP1+
6	CP1-
7	AIN0.0
8	AIN0.1
9	DAC0
10	DAC1
11	AIN0.6
12	AIN0.7

SILICON LABORATORIES

6.6. PORT I/O Connectors (J12 - J19)

In addition to all port I/O signals being routed to the 96-pin expansion connector, each of the eight parallel ports of the C8051F020 has its own 10-pin header connector. Each connector provides a pin for the corresponding port pins 0-7, +3.3 VDC and digital ground. Table 5 defines the pins for the port connectors. The same pin-out order is used for all of the port connectors.

Table 5. J12- J19 Port Connector Pin Descriptions

Pin #	Description
1	Pn.0
2	Pn.1
3	Pn.2
4	Pn.3
5	Pn.4
6	Pn.5
7	Pn.6
8	Pn.7
9	+3 VD (+3.3 VDC)
10	GND (Ground)

6.7. VDD Monitor Disable (J23)

The VDD Monitor of the C8051F020 may be disabled by moving the shorting block on J23 from pins 1-2 to pins 2-3, as shown in Figure 3.

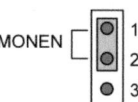

Figure 3. VDD Monitor Hardware Setup

6.8. VREF Connector (J22)

The VREF connector (J22) can be used to connect the VREF (Voltage Reference) output of the C8051F020 to any (or all) of its voltage reference inputs. Install shorting blocks on J22 in the following manner:

1-2 to connect VREF to VREFD
3-4 to connect VREF to VREF0
5-6 to connect VREF to VREF1

SILICON LABORATORIES

6.9. Expansion I/O Connector (J24)

The 96-pin expansion I/O connector J24 is used to connect daughter boards to the main target board. J24 provides access to many C8051F020 signal pins. Pins for +3 V, digital ground, analog ground and the unregulated power supply (VUNREG) are also available. The VUNREG pin is connected directly to the unregulated +V pin of the P1 power connector. See Table 6 for a complete list of pins available at J24.

The J24 socket connector is manufactured by Hirose Electronic Co. Ltd, part number PCN13-96S-2.54DS, Digi-Key part number H7096-ND. The corresponding plug connector is also manufactured by Hirose Electronic Co. Ltd, part number PCN10-96P-2.54DS, Digi-Key part number H5096-ND.

Table 6. J24 Pin Descriptions

Pin #	Description	Pin #	Description	Pin #	Description
A-1	+3 VD2 (+3.3 VDC)	B-1	DGND (Digital Gnd)	C-1	XTAL1
A-2	MONEN	B-2	P1.7	C-2	P1.6
A-3	P1.5	B-3	P1.4	C-3	P1.3
A-4	P1.2	B-4	P1.1	C-4	P1.0
A-5	P2.7	B-5	P2.6	C-5	P2.5
A-6	P2.4	B-6	P2.3	C-6	P2.2
A-7	P2.1	B-7	P2.0	C-7	P3.7
A-8	P3.6	B-8	P3.5	C-8	P3.4
A-9	P3.3	B-9	P3.2	C-9	P3.1
A-10	P3.0	B-10	P0.7	C-10	P0.6
A-11	P0.5	B-11	P0.4	C-11	P0.3
A-12	P0.2	B-12	P0.1	C-12	P0.0
A-13	P7.7	B-13	P7.6	C-13	P7.5
A-14	P7.4	B-14	P7.3	C-14	P7.2
A-15	P7.1	B-15	P7.0	C-15	P6.7
A-16	P6.6	B-16	P6.5	C-16	P6.4
A-17	P6.3	B-17	P6.2	C-17	P6.1
A-18	P6.0	B-18	P5.7	C-18	P5.6
A-19	P5.5	B-19	P5.4	C-19	P5.3
A-20	P5.2	B-20	P5.1	C-20	P5.0
A-21	P4.7	B-21	P4.6	C-21	P4.5
A-22	P4.4	B-22	P4.3	C-22	P4.2
A-23	P4.1	B-23	P4.0	C-23	TMS
A-24	TCK	B-24	TDI	C-24	TDO
A-25	/RST	B-25	DGND (Digital Gnd)	C-25	VUNREG
A-26	AGND (Analog Gnd)	B-26	DAC1	C-26	DAC0
A-27	CP1-	B-27	CP1+	C-27	CP0-
A-28	CP0+	B-28	VREF	C-28	VREFD
A-29	VREF0	B-29	VREF1	C-29	AIN0.7
A-30	AIN0.6	B-30	AIN0.5	C-30	AIN0.4
A-31	AIN0.3	B-31	AIN0.2	C-31	AIN0.1
A-32	AIN0.0	B-32	AGND (Analog Gnd)	C-32	AV+ (+3.3 VDC Analog)

7. Schematic

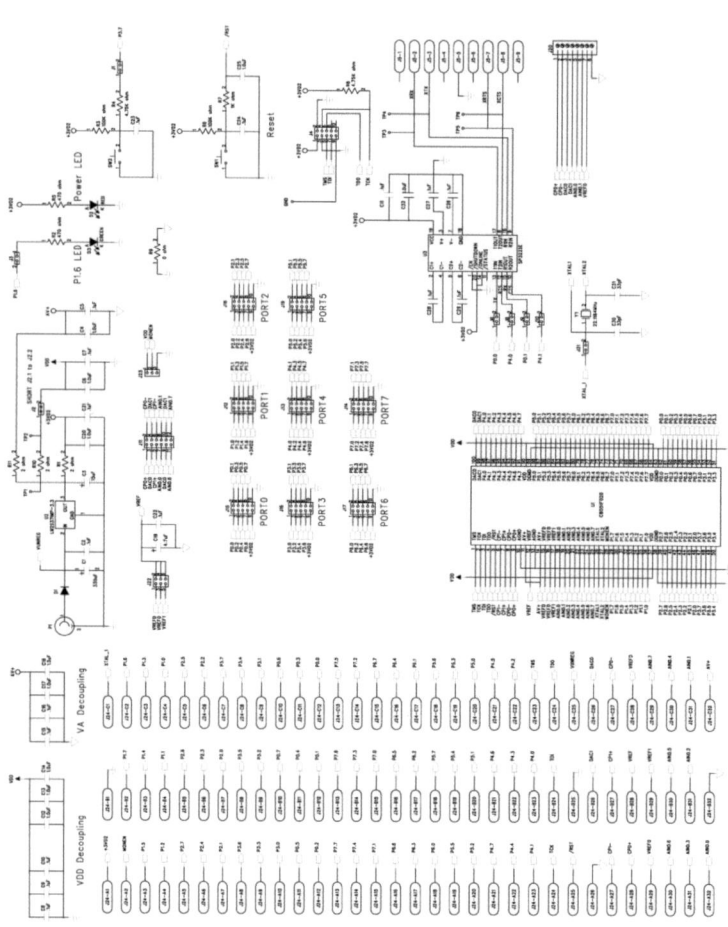

Figure 4. C8051F020 Target Board Schematic

XXVI

DOCUMENT CHANGE LIST

Revision 0.4 to Revision 0.5

- Section 1, added USB Debug Adapter and USB Cable.
- Section 2, changed name from "Hardware Setup" to "Hardware Setup using an EC2 Serial Adapter".
- Section 2, added 2 Notes bullets.
- Section 2, removed Note from bottom of page.
- Added Section 3, "Hardware Setup using a USB Debug Adapter".
- Section 5.4.2, changed step 2 to include new instructions.
- Section 7, J4, changed "Serial Adapter" to "Debug Adapter".
- Target Board DEBUG Interface Section, added USB Debug Adapter.
- DEBUG Connector Pin Descriptions Table, changed pin 4 to C2D.
- Changed "jumper" to "header".
- EC2 Serial Adapter section, added EC2 to the section title, table title and figure title.
- EC2 Serial Adapter section, changed "JTAG" to "DEBUG".
- Added "USB Debug Adapter" section.
- Added J8 and J10 to the figure in the Target Board section.
- Added J8 and J10 to the connector list.

Revision 0.5 to Revision 0.6

- Removed EC2 Serial Adapter from Kit Contents.
- Removed Section 2. Hardware Setup using an EC2 Serial Adapter. See RS232 Serial Adapter (EC2) User's Guide.
- Removed Section 8. EC2 Serial Adapter. See RS232 Serial Adapter (EC2) User's Guide.
- Removed Section 9. USB Debug Adapter. See USB Debug Adapter User's Guide.

C8051F02x-DK

CONTACT INFORMATION

Silicon Laboratories Inc.
4635 Boston Lane
Austin, TX 78735
Tel: 1+(512) 416-8500
Fax: 1+(512) 416-9669
Toll Free: 1+(877) 444-3032

Email: MCUinfo@silabs.com
Internet: www.silabs.com

Anhang 3: Programmcode „Blinky stripped"

```c
// C. Hahlweg, Nov 2011: modified and stripped down from
//-------------------------------------------------------------------------------
// F02x_Blinky.c / example collection Silabs
//-------------------------------------------------------------------------------
//
// Flashes the green LED on the C8051F020 target board about five times
// a second using the interrupt handler for Timer3.
// Target: C8051F02x
//
// Tool chain: KEIL Eval 'c'
//
//
//-------------------------------------------------------------------------------
// Includes
//-------------------------------------------------------------------------------
#include <c8051f020.h>                 // SFR declarations


//-------------------------------------------------------------------------------
// 16-bit SFR Definitions for 'F02x

// 8-bit definitions see <c8051f020.h>
//-------------------------------------------------------------------------------
// Timer 3 : see Datasheet 8051 F020: 16-bit time with auto-relaod only.
// Note: Timers 2 and 4 could be used as well, but are saved for their flexibility


sfr16 TMR3RL    = 0x92;                // Timer3 reload value
sfr16 TMR3      = 0x94;                // Timer3 counter

//-------------------------------------------------------------------------------
// Global CONSTANTS
//-------------------------------------------------------------------------------

#define SYSCLK 2000000                 // 2MHz for OSCICN.[1:0] = 00
                                                     // approximate

SYSCLK frequency in Hz

sbit  LED = P1^6;                      // green LED: '1' = ON; '0' = OFF
char  oscmask = 0x00;                           // OSCICN.[1:0] = 00 2MHz //
01 4MHz 10 8MHz 11 16MHz


//-------------------------------------------------------------------------------
// Function PROTOTYPES
//-------------------------------------------------------------------------------
void PORT_Init (void);
void Timer3_Init (int counts);
void Timer3_ISR (void);

//-------------------------------------------------------------------------------
// MAIN Routine
//-------------------------------------------------------------------------------
void main (void) {

    // disable watchdog timer
    WDTCN = 0xde;
    WDTCN = 0xad;
    // set internal oscillator frequency
    OSCICN &= 0xfc;
    OSCICN |= oscmask;
    //fact = fact << oscmask;
```

```c
    PORT_Init ();
    Timer3_Init ( SYSCLK / 12 / 1);          // Init Timer3 to generate interrupts
                                             // at a 10Hz rate. for 2 MHz


    EA = 1;                                                                      //
enable global interrupts

    while (1) {                              // spin forever
    }
}

//-------------------------------------------------------------------------------
// PORT_Init
//-------------------------------------------------------------------------------
//
// Configure the Crossbar and GPIO ports
//
void PORT_Init (void)
{
    XBR2    = 0x40;                     // Enable crossbar and weak pull-ups
    P1MDOUT |= 0x40;                    // enable P1.6 (LED) as push-pull output
}


//-------------------------------------------------------------------------------
// Timer3_Init
//-------------------------------------------------------------------------------
//
// Configure Timer3 to auto-reload and generate an interrupt at interval
// specified by <counts> using SYSCLK/12 as its time base.
//
void Timer3_Init (int counts)
{
    TMR3CN = 0x00;                     // Stop Timer3; Clear TF3;
                                       // use SYSCLK/12 as timebase
    TMR3RL  = 0xffff-counts;               // Init reload values
    TMR3    = 0xffff;                  // set to reload immediately
    EIE2   |= 0x01;                    // enable Timer3 interrupts
    TMR3CN |= 0x04;                    // start Timer3
}

//-------------------------------------------------------------------------------
// Interrupt Service Routines
//-------------------------------------------------------------------------------


//-------------------------------------------------------------------------------
// Timer3_ISR
//-------------------------------------------------------------------------------
// This routine changes the state of the LED whenever Timer3 overflows.
//
void Timer3_ISR (void) interrupt 14
{
    TMR3CN &= ~(0x80);                 // clear TF3
    LED = ~LED;                        // change state of LED
}                                      // results in half the interrupt frequency
```

Anhang 4: Programmcode „Uhr"

```c
// C. Hahlweg, Nov 2011: modified and stripped down from
//-----------------------------------------------------------------------------
// F02x_Blinky.c / example collection Silabs
//-----------------------------------------------------------------------------
//
// Flashes the green LED on the C8051F020 target board about five times
// a second using the interrupt handler for Timer3.
// Target: C8051F02x
//
// Tool chain: KEIL Eval 'c'
//-----------------------------------------------------------------------------
// Includes
//-----------------------------------------------------------------------------
#include <c8051f020.h>                    // SFR declarations

//-----------------------------------------------------------------------------
// 16-bit SFR Definitions for 'F02x

// 8-bit definitions see <c8051f020.h>
//-----------------------------------------------------------------------------
// Timer 3 : see Datasheet 8051 F020: 16-bit time with auto-relaod only.
// Note: Timers 2 and 4 could be used as well, but are saved for their flexibility


sfr16 TMR3RL    = 0x92;                   // Timer3 reload value
sfr16 TMR3      = 0x94;                   // Timer3 counter


//-----------------------------------------------------------------------------
// Global CONSTANTS
//-----------------------------------------------------------------------------

#define SYSCLK 2000000                       // 2MHz for OSCICN.[1:0] = 00
                                                        // approximate
SYSCLK frequency in Hz

//sbit  LED = P1^6;                          // green LED: '1' = ON; '0' = OFF
char  oscmask = 0x00;                             // OSCICN.[1:0] = 00 2MHz //
01 4MHz 10 8MHz 11 16MHz

char  wert_sec_10  = 0x00;                      // 1/10 sec
char  wert_sec_1   = 0x00;             // 1 sec
char  wert_sec_6   = 0x00;            // 60 sec
char  wert_min_1   = 0x00;            // 1 -9 min

char  SSeg[10] = {1+2+4+8+16+32,    //0
                  2+4,              //1
                    1+4+8+32+64,      //2
                    1+2+4+8+64,       //3
                    2+4+16+64,        //4
                    1+2+8+16+64,      //5
                    1+2+8+16+32+64+128, //6
                    2+4+8,            //7
                    255 -128,         //8
                    1+2+4+8+16+64     //9
                    };

//-----------------------------------------------------------------------------
// Function PROTOTYPES
//-----------------------------------------------------------------------------
void PORT_Init (void);
void Timer3_Init (int counts);
```

```c
void Timer3_ISR (void);
//void OSCILLATOR_Init (void);

//-----------------------------------------------------------------------------
// MAIN Routine
//-----------------------------------------------------------------------------
void main (void) {

    // disable watchdog timer
    WDTCN = 0xde;
    WDTCN = 0xad;
    // set internal oscillator frequency
    OSCICN &= 0xfc;
    OSCICN |= oscmask;
    //OSCILLATOR_Init ();

    PORT_Init ();
    Timer3_Init ( SYSCLK / 12 / 10);        // Init Timer3 to generate interrupts
                                            // at a 10Hz rate. for 2 MHz

    EA = 1;                                                             //
enable global interrupts

    while (1) {                          // spin forever
    }
}

//-----------------------------------------------------------------------------
// PORT_Init
//-----------------------------------------------------------------------------
//
// Configure the Crossbar and GPIO ports
//
void PORT_Init (void)
{
    XBR2    = 0x40;                  // Enable crossbar and weak pull-ups
    //P1MDOUT |= 0x40;               // enable P1.6 (LED) as push-pull output
    P0MDOUT = 0xff;
    P1MDOUT = 0xff;
    P2MDOUT = 0xff;
}

//-----------------------------------------------------------------------------
// Timer3_Init
//-----------------------------------------------------------------------------
//
// Configure Timer3 to auto-reload and generate an interrupt at interval
// specified by <counts> using SYSCLK/12 as its time base.
//
void Timer3_Init (int counts)
{
    TMR3CN = 0x00;                      // Stop Timer3; Clear TF3;
                                        // use SYSCLK/12 as timebase
    TMR3RL  = 0xffff-counts;            // Init reload values
    TMR3    = 0xffff;                   // set to reload immediately
    EIE2   |= 0x01;                     // enable Timer3 interrupts
    TMR3CN |= 0x04;                     // start Timer3
}
// Init ext. Osc._------------------------------------------
// -------------------------------------------------------
/*void OSCILLATOR_Init (void)
{
    int i;                          // delay counter
```

```
    OSCXCN = 0x67;                          // start external oscillator with
                                            // 22.1184MHz crystal

    for (i=0; i < 256; i++) ;               // wait for oscillator to start

    while (!(OSCXCN & 0x80)) ;              // Wait for crystal osc. to settle

    OSCICN = 0x88;                          // select external oscillator as SYSCLK
                                            // source and enable missing clock
                                            // detector

} */

//-------------------------------------------------------------------------------
// Interrupt Service Routines
//-------------------------------------------------------------------------------

//-------------------------------------------------------------------------------
// Timer3_ISR
//-------------------------------------------------------------------------------
// This routine changes the state of the LED whenever Timer3 overflows.
//
void Timer3_ISR (void) interrupt 14
{
    TMR3CN &= ~(0x80);                      // clear TF3
    //LED = ~LED;                           // change state of LED

    P0 = SSeg[wert_sec_1];
    P1 = SSeg[wert_sec_6];
    P2 = SSeg[wert_min_1];

    wert_sec_10 = wert_sec_10 +1;
    if(wert_sec_10== 0x0a)
    {
     wert_sec_10 = 0x00;
     wert_sec_1 = wert_sec_1 + 1;
        if(wert_sec_1 ==0x0a)               // 0x0a = 10Dec
        {
          wert_sec_1 = 0x00;
          wert_sec_6 = wert_sec_6 + 1;
          if(wert_sec_6 == 0x06)
          {
            wert_sec_6 = 0x00;
              wert_min_1 = wert_min_1 +1;
              if(wert_min_1 == 0x0a) wert_min_1 = 0x00;


          }
        };
    };

    //P2 = wert;
}                                           // results in half the interrupt frequency
```